© 2024 by FAISAL JAMIL. All rights reserved.

Title: "Rising Sea Levels: A Threat to Coastal Cities"

This book, along with its contents encompassing text, illustrations, images, diagrams, and other creative elements, is the exclusive property of FAISAL JAMIL and is safeguarded by copyright law.

FAISAL JAMIL asserts full ownership and retains all rights to this book. No part of this publication may be reproduced, distributed, or transmitted in any form or by any means, such as photocopying, recording, or electronic methods, without prior written consent from the copyright holder. Brief quotations in critical reviews and certain noncommercial uses permitted by copyright law are exceptions.

This copyright notice applies to all editions, formats, and translations of the book, whether in print, digital, or any other medium or technology existing now or developed in the future. Unauthorized use or infringement may result in legal action and pursuit of remedies under applicable copyright laws.

While efforts have been made to ensure accuracy and reliability, FAISAL JAMIL does not guarantee the completeness or suitability of the information. Readers are responsible for evaluating and using the content judiciously.

FAISAL JAMIL reserves the right to make changes, updates, or corrections to the book without prior notice. Inclusion of third-party materials or references does not imply endorsement or affiliation unless used under fair use principles or with proper permissions and attributions.

For permissions, inquiries, or requests regarding the book's use, please contact FAISAL JAMIL through official channels listed on their Amazon author page or provided email address.

This comprehensive copyright notice serves to protect FAISAL JAMIL'S intellectual property rights, maintain content control, and inform users about associated restrictions and permissions.

Warm regards,

FAISAL JAMIL

I Always Give's Free Copies Need Your Feedback And Reviews Keeps In Touch!

http://www.amazon.com/author/faisal.jamil

Email: faisaljamilauthor@gmail.com

About the author

Certainly! Faisal Jamil is a multifaceted individual with a diverse set of skills and experiences. With a strong foundation in computer knowledge since childhood, he has developed a deep understanding of technology that informs his work as a content writer. Faisal also possesses digital skills, which further enhance his abilities in various digital platforms and technologies.

Beyond his professional endeavors, Faisal Jamil has also excelled in the martial arts, particularly Shotokan Karate, where he achieved the prestigious rank of first Dan black belt. This achievement speaks to his dedication, discipline, and commitment to personal growth and mastery.

In his professional life, Faisal Jamil has carved out a successful career in sales management within the Fast Moving Consumer Goods (FMCG) sector. His roles in various FMCG companies have honed his skills in strategic planning, team leadership, and business development. Faisal's ability to drive sales and achieve targets has been instrumental in his career progression, showcasing his talent for identifying opportunities and delivering results.

Faisal Jamil is also deeply interested in business investment strategies, planning, and execution. His understanding of these areas has been key to his success in the business world, allowing him to make informed decisions and implement effective strategies. His ability to navigate the complexities of investment planning and execution has set him apart as a strategic thinker and a valuable asset in any business endeavor.

Overall, Faisal Jamil is a dynamic individual who combines his passion for technology, martial arts, sales management, digital skills, and business investment strategies to achieve success in diverse fields. His journey is a testament to his versatility, resilience, and continuous pursuit of excellence.

Yours Sincerely

FAISAL JAMIL

I Always Give's Free Copies Need Your Feedback And Reviews Keeps In Touch!

https://www.amazon.com/author/faisal.jamil

Email: faisaljamilauthor@gmail.com

RISING SEA LEVELS A THREAT TO COASTAL CITIES

Table of Content

Preface -- 8

Introduction --- 10

Chapter 1: The Calm Before the Storm ----------------------- 16

Chapter 2: Unseen Forces -- 20

Chapter 3: Early Warnings --------------------------------------- 24

Chapter 4: The Economic Impact ------------------------------- 29

Chapter 5: The Human Cost -------------------------------------- 34

Chapter 6: The Role of Governments -------------------------- 39

Chapter 7: Technological Innovations ------------------------- 46

Chapter 8: Coastal Ecosystems at Risk ------------------------ 53

Chapter 9: Voices from the Frontlines ------------------------ 60

Chapter 10: The Global Perspective ---------------------------- 66

Chapter 11: Urban Planning for a Watery Future ---------- 74

Chapter 12: The Power of Awareness --------------------------- 83

Chapter 13: Adaptation vs. Mitigation------------------------- 91

Chapter 14: The Insurance Dilemma -------------------------- 101

Chapter 15: Legal and Ethical Considerations -------------- 110

Chapter 16: Cultural Heritage at Risk ------------------------ 119

Chapter 17: Youth Voices and Future Leaders --------------129

Chapter 18: The Role of Industry -----------------------------135

Chapter 19: Hope on the Horizon -----------------------------146

Chapter 20: A Call to Action ----------------------------------154

Preface

In the early mornings, as the sun begins to rise over the serene waters of Seaside, it casts a golden glow that promises another beautiful day. Yet beneath this tranquil surface lies a growing threat that has the potential to reshape not only this quaint coastal town but countless others around the world. The rising sea levels, driven by climate change, pose a significant and imminent danger to coastal cities, their ecosystems, and the millions of people who call these places home.

The journey to writing "Rising Sea Levels: A Threat to Coastal Cities" began with a simple observation of the subtle changes occurring in coastal regions. Unusual tides, persistent coastal erosion, and increasing instances of flooding became impossible to ignore. As I delved deeper into the science and stories behind these phenomena, it became clear that rising sea levels are not just an environmental issue but a complex, multifaceted challenge that intersects with economics, politics, culture, and human resilience.

This book is a comprehensive exploration of the causes, impacts, and responses to rising sea levels. It is structured to provide a holistic understanding of the issue, beginning with the foundational science and extending through the personal and societal implications. Each chapter aims to shed light on different aspects of this global crisis, from the historical precedents that offer valuable lessons to the innovative solutions that provide hope for the future.

Throughout this journey, I have had the privilege of engaging with leading scientists, passionate activists, resilient community leaders, and everyday citizens who are on the frontlines of this battle. Their insights, experiences, and unwavering dedication have enriched this book and inspired its narrative. These voices are crucial in understanding the human dimension of rising sea levels and the collective action required to address it.

"Rising Sea Levels: A Threat to Coastal Cities" is not just a presentation of facts and figures. It is a call to action. The challenges we face are immense, but so too are the opportunities for innovation, resilience, and cooperation. This book aims to empower readers with knowledge and inspire them to take part in the global effort to protect our coastal cities and preserve our planet for future generations.

As you turn these pages, I hope you will be moved by the stories of those who are already making a difference and motivated to join them. Whether you are an individual looking to make more sustainable choices, a community leader seeking to implement local adaptation strategies, or a policymaker shaping the future, your actions matter. Together, we can rise to meet this challenge and ensure a safer, more resilient future for all.

Thank you for embarking on this journey with me.

Sincerely,

FAISAL JAMIL

INTRODUCTION

As dawn breaks over the coastal town of Seaside, its residents awaken to the soothing sound of waves gently lapping against the shore. This daily rhythm, a source of comfort and livelihood for generations, is now under threat. Subtle but significant changes in the natural environment signal an urgent and growing crisis—rising sea levels. What was once a distant concern has now become a tangible and imminent danger for coastal cities around the world.

The issue of rising sea levels is a complex and multifaceted challenge. It is not merely an environmental problem but one that intertwines with economic stability, social equity, cultural heritage, and political governance. As ice caps melt and oceans warm, the implications are far-reaching, affecting ecosystems, infrastructure, and human lives in profound ways.

This book, "Rising Sea Levels: A Threat to Coastal Cities," is a comprehensive examination of this global phenomenon. It seeks to unravel the intricate web of causes, impacts, and responses to rising sea levels. Through a blend of scientific analysis, historical context, personal narratives, and case studies, we will explore the various dimensions of this crisis and the efforts being made to combat it.

The Scope of the Problem

At the heart of rising sea levels is the undeniable reality of climate change. Human activities, particularly the burning of fossil fuels and deforestation, have significantly increased greenhouse gas emissions, leading to global warming. This warming has two primary effects on sea levels: the thermal expansion of seawater and the melting of polar ice caps and glaciers. Together, these processes contribute to the gradual but relentless rise of ocean waters.

The consequences of rising sea levels are manifold. Coastal erosion accelerates, swallowing beaches and threatening coastal properties. Saltwater intrusion contaminates freshwater supplies, affecting agriculture and drinking water. More frequent and severe flooding endangers lives, disrupts economies, and strains infrastructure. The compounded effects can lead to displacement and migration, creating new challenges for communities and nations.

Historical and Contemporary Perspectives

Understanding the full scope of rising sea levels requires looking both to the past and the present. Throughout history, civilizations have faced the consequences of changing sea levels. Ancient coastal settlements that once thrived now lie submerged beneath the ocean, serving as poignant reminders of nature's power. In modern times, advancements in science and technology have allowed us to predict and monitor these changes more accurately, but the challenge remains daunting.

A Global and Local Crisis

While rising sea levels are a global phenomenon, their impacts are felt most acutely at the local level. Coastal cities across the world, from New York to Jakarta, face unique challenges based on their geography, economy, and infrastructure. This book will highlight diverse case studies from various regions, illustrating the shared and distinct experiences of these communities. By examining these examples, we gain insights into the different strategies employed to mitigate and adapt to the rising waters.

Innovative Solutions and Resilient Communities

Despite the enormity of the challenge, there is hope. Innovative technologies and forward-thinking urban planning offer potential solutions to enhance resilience. From building sea walls and designing floating cities to restoring natural barriers like mangroves and coral reefs, humanity is leveraging creativity and science to adapt to changing realities. Equally important are the grassroots efforts and community-led initiatives that demonstrate the power of collective action and local knowledge.

The Call to Action

The final chapters of this book focus on the crucial need for action at all levels—individual, community, national, and international. It emphasizes the role of governments in enacting policies and regulations, the responsibility of industries in reducing their carbon footprint, and the importance of public awareness and education. Ultimately, it is a collective effort that will determine our ability to navigate and mitigate the impacts of rising sea levels.

"Rising Sea Levels: A Threat to Coastal Cities" is not just an exploration of a pressing issue but a call to action. It urges readers to understand the gravity of the situation, recognize the interconnectedness of our world, and engage in solutions that promote sustainability and resilience. As we journey through the pages of this book, let us be inspired by the stories of those who are making a difference and be motivated to join them in safeguarding our coastal cities and the planet we all share.

Welcome to this exploration of rising sea levels, a journey that challenges us to rise to the occasion and create a more resilient future for generations to come.

Sincerely,

FAISAL JAMIL

Chapter 1
The Calm Before the Storm

Seaside: A Town in Harmony

Seaside, a picturesque coastal town nestled between rolling hills and the expansive ocean, has always been a place of tranquility and beauty. With its white sandy beaches, charming boardwalks, and vibrant marine life, the town is a haven for both its residents and visitors. For generations, Seaside's inhabitants have lived in harmony with the ocean, their lives deeply intertwined with the rhythms of the sea.

The town's history is rich with tales of seafaring adventures, prosperous fishing industries, and festivals that celebrate the bounty of the ocean. Every year, Seaside hosts the Sea Festival, a week-long event where the community comes together to honor their maritime heritage with parades, seafood feasts, and boat races. The festival is a testament to the town's enduring connection to the ocean.

Economically, Seaside thrives on its coastal resources. The fishing industry provides a livelihood for many families, with generations of fishermen passing down their knowledge and skills. Tourism is another vital part of the local economy, as visitors flock to the town to enjoy its scenic beauty, recreational activities, and historic landmarks.

Subtle Changes

Despite the idyllic setting, subtle changes begin to emerge, signaling a shift in the town's delicate balance with the ocean. Residents notice unusual tides that seem higher than usual, encroaching further onto the shore and sometimes flooding the lower streets. The once stable coastline starts to show signs of erosion, with chunks of land gradually disappearing into the sea.

One of the first to notice these changes is Tom, a local fisherman whose family has lived in Seaside for generations. Tom's father and grandfather taught him the secrets of the sea, and he's spent his life understanding its patterns and moods. Recently, he's observed that the fish are behaving differently, migrating earlier and to different areas. He also notes that his boat, which he's meticulously maintained, seems to be encountering more debris and unusual currents.

Sophie, a marine biologist who moved to Seaside a few years ago, conducts regular studies of the coastal ecosystem. She documents the increasing erosion and unusual tide patterns, realizing that these are early signs of a larger issue. Sophie tries to raise awareness among the residents and local authorities, but her warnings are often met with skepticism or indifference. Many believe these changes are just part of natural cycles they've seen before.

Rising Concerns

As the subtle changes become more frequent and pronounced, concerns start to grow among the townspeople. Mary's beachfront cafe, a beloved spot for both locals and tourists, experiences occasional flooding that damages her property and disrupts business. She installs sandbags and raises her patio, but these measures seem increasingly inadequate.

The town's children, too, notice the changes. During a school field trip to the beach, they find that their favorite tide pools are disappearing. Their teacher, Mr. Johnson, uses this as an opportunity to teach them about climate change and its potential impacts, planting seeds of awareness in young minds.

Community meetings begin to address these issues, but there's a division among residents. Some, like Tom and Mary, advocate for immediate action, while others are hesitant to acknowledge the problem, fearing the economic implications. The local government, constrained by limited resources, struggles to balance the need for infrastructure improvements with the demands of other pressing issues.

The Looming Crisis

Unbeknownst to the townspeople, Seaside is on the brink of a significant transformation. Scientists like Sophie understand that these early signs—unusual tides, coastal erosion, and occasional flooding—are foreshadowing a looming crisis. The global phenomenon of rising sea levels, driven by climate change, is beginning to impact Seaside in ways that will challenge its very existence.

The chapter closes with a poignant scene: Tom, standing on the shore at dawn, gazing out at the ocean. The sea, which has always been a source of life and livelihood, now seems unpredictable and ominous. As the first rays of sunlight reflect off the waves, Tom resolves to do whatever it takes to protect his home and way of life, even as he grapples with the uncertainty of what lies ahead.

In "The Calm Before the Storm," the stage is set for a dramatic and urgent narrative that will explore the challenges, conflicts, and resilience of a community facing the inexorable rise of the sea.

Chapter 2
Unseen Forces

Introduction to the Science

In the serene setting of Seaside, the subtle changes in the environment are just the tip of the iceberg. To understand the full scope of the looming crisis, it's essential to delve into the science behind rising sea levels. This chapter explores the complex interplay of natural and human-induced factors driving the phenomenon, providing a comprehensive understanding of the forces at work.

Melting Ice Caps and Glaciers

One of the most significant contributors to rising sea levels is the melting of ice caps and glaciers. In the polar regions, vast ice sheets in Greenland and Antarctica hold enormous amounts of frozen water. As global temperatures rise due to climate change, these ice sheets are melting at an accelerating rate.

Dr. Emily White, a glaciologist who has spent years studying the polar ice caps, explains, "The ice sheets are losing mass at an unprecedented rate. In Greenland alone, we see billions of tons of ice melting each year, contributing significantly to global sea level rise." She shows satellite images that starkly illustrate the shrinking ice caps, a visual testament to the scale of the problem.

Glaciers in mountainous regions are also retreating. The Himalayas, the Alps, and the Andes are all experiencing significant ice loss. These glaciers have traditionally acted as natural reservoirs, releasing water gradually. Their rapid melting disrupts this balance, leading to increased runoff and contributing to sea level rise.

Thermal Expansion of Water

Another critical factor is the thermal expansion of seawater. As the Earth's atmosphere warms, so do the oceans. Warmer water occupies more volume than colder water, leading to an increase in sea level. This process, while seemingly subtle, has a profound impact over large scales.

Oceanographer Dr. James Liu explains, "Thermal expansion accounts for about half of the observed sea level rise over the past century. Even small increases in temperature can result in significant volume changes when applied to the vast expanse of the world's oceans."

Dr. Liu's research involves monitoring ocean temperatures and currents. His data indicates a clear upward trend in ocean temperatures, correlating strongly with global warming patterns. This thermal expansion is a silent but powerful driver of rising sea levels.

Human Activities and Climate Change

Human activities play a crucial role in exacerbating climate change, which in turn drives sea level rise. The burning of fossil fuels, deforestation, and industrial processes release vast amounts of greenhouse gases into the atmosphere. Carbon dioxide (CO_2), methane (CH_4), and nitrous oxide (N_2O) trap heat, leading to global warming.

Climate scientist Dr. Maria Gonzalez discusses the human impact, "The increase in greenhouse gas emissions since the industrial revolution has created an enhanced greenhouse effect. This effect is causing the Earth's temperature to rise, which leads to the melting of ice caps, glaciers, and the thermal expansion of oceans."

Dr. Gonzalez's models predict that if current emission trends continue, sea levels could rise by up to a meter by the end of the century. Such a rise would have catastrophic implications for coastal communities worldwide.

Interviews with Scientists and Experts

To paint a clearer picture of the impending danger, this chapter includes interviews with several scientists and experts who provide diverse perspectives on the issue.

Dr. Emily White (Glaciologist):

"We are witnessing changes in the polar ice sheets that we never thought possible. The rate of ice loss is accelerating, and the potential for significant sea level rise is very real. We need immediate global action to mitigate these effects."

Dr. James Liu (Oceanographer):

"Thermal expansion is often overlooked, but it's a major player in sea level rise. Our oceans are absorbing a lot of the excess heat from global warming, and this has a direct impact on sea levels."

Dr. Maria Gonzalez (Climate Scientist):

"The human contribution to climate change is undeniable. We must reduce greenhouse gas emissions and transition to renewable energy sources if we hope to slow the rate of sea level rise."

The Impending Danger

The chapter concludes with a synthesis of the scientific findings. The combination of melting ice caps, thermal expansion, and human-induced climate change presents a clear and present danger to coastal communities like Seaside. The subtle changes observed by residents are just the beginning. Without significant global efforts to address climate change, the effects on sea levels will continue to intensify, leading to more frequent and severe flooding, coastal erosion, and loss of habitable land.

In "Unseen Forces," the science behind rising sea levels is brought to the forefront, emphasizing the urgency of the situation. The interviews with experts underscore the need for immediate action and global cooperation to mitigate the impending crisis. The chapter sets the stage for understanding the broader implications of rising sea levels and the need for resilient solutions to protect coastal communities.

Chapter 3
Early Warnings

Introduction to Historical Context

Understanding the threat of rising sea levels requires a look back into history. This chapter examines historical events and past warnings about sea level rise, drawing parallels between ancient civilizations that vanished due to rising waters and modern-day predictions. These early warnings emphasize the importance of recognizing and responding to the signs of environmental change before it's too late.

Ancient Civilizations and Rising Waters

Throughout history, numerous civilizations have faced the consequences of rising sea levels. These events provide valuable lessons about the vulnerability of human settlements to environmental changes.

The Lost City of Dwarka:

According to Indian mythology, Dwarka was a magnificent city ruled by Lord Krishna. Archaeological evidence suggests that an ancient city did exist off the coast of present-day Gujarat, which was submerged due to rising sea levels around 1500 BCE. The ruins of this city, now underwater, are a testament to the destructive power of the ocean.

The Minoan Civilization:

The Minoan civilization on the island of Crete flourished between 3000 and 1450 BCE. Archaeologists believe that a combination of seismic activity and rising sea levels contributed to the decline of this advanced society. The eruption of the Thera volcano and subsequent tsunamis likely exacerbated the situation, highlighting the complex interplay between geological events and sea level changes.

Doggerland:

Once a land bridge connecting Great Britain to mainland Europe, Doggerland was inhabited by Mesolithic humans around 10,000 BCE. Rising sea levels caused by melting glaciers gradually submerged this area, forcing its inhabitants to migrate. The story of Doggerland serves as a reminder of how climate-induced sea level rise can reshape human geography.

Modern-Day Predictions

While ancient events offer historical context, modern science provides precise predictions about future sea level rise. These predictions, based on extensive research and advanced modeling techniques, offer early warnings that should not be ignored.

Intergovernmental Panel on Climate Change (IPCC):

The IPCC's reports are among the most comprehensive sources of information on climate change and its impacts. The 2021 Sixth Assessment Report predicts that global sea levels could rise by up to 1.1 meters by 2100 if greenhouse gas emissions continue unabated. Such a rise would inundate coastal cities and displace millions of people.

National Aeronautics and Space Administration (NASA):

NASA's satellite data and climate models provide detailed insights into sea level trends. According to their projections, sea levels are rising at an accelerated rate, with an average increase of about 3.3 millimeters per year. This data underscores the urgency of addressing the root causes of climate change.

National Oceanic and Atmospheric Administration (NOAA)

NOAA's sea level rise scenarios for the United States predict significant impacts on coastal regions. By 2100, sea levels along the U.S. coastline could rise by 0.3 to 2.5 meters, depending on future emissions. These predictions highlight the need for adaptive measures to protect vulnerable communities.

Past Warnings and Modern Lessons

Historical and modern warnings offer crucial lessons about the importance of early action and proactive measures to address rising sea levels.

The Netherlands:

The Dutch have a long history of battling the sea. The catastrophic North Sea Flood of 1953, which claimed over 1,800 lives, prompted the Netherlands to implement the Delta Works project. This extensive system of dams, sluices, locks, dikes, and storm surge barriers has successfully protected the country from further flooding. The Dutch experience underscores the importance of investing in robust infrastructure to mitigate the impacts of rising sea levels.

Hurricane Katrina (2005):

The devastation wrought by Hurricane Katrina on New Orleans serves as a stark warning about the vulnerability of coastal cities to extreme weather events exacerbated by rising sea levels. The failure of levees and inadequate preparedness led to widespread destruction and loss of life. Katrina's aftermath highlighted the need for comprehensive disaster planning and resilient infrastructure.

Pacific Island Nations:

Small island nations in the Pacific, such as Tuvalu and Kiribati, are on the frontlines of sea level rise. These countries have been vocal about the existential threat posed by rising waters, calling for global action to combat climate change. Their plight serves as a powerful reminder of the urgent need for international cooperation and support for vulnerable regions.

The Imperative of Heeding Early Warnings

The chapter concludes by emphasizing the critical importance of heeding early warnings. Both historical events and modern predictions highlight the devastating consequences of failing to recognize and respond to the signs of rising sea levels. As Seaside begins to experience subtle changes, the lessons from the past and the insights from scientific predictions provide a clear message: proactive measures and early action are essential to mitigate the impacts of this global challenge.

In "Early Warnings," the narrative weaves together historical accounts and modern scientific predictions, underscoring the importance of learning from the past to inform the future. The chapter sets the stage for exploring the specific economic, social, and environmental impacts of rising sea levels on coastal communities like Seaside, reinforcing the urgency of addressing this pressing issue.

Chapter 4
The Economic Impact

Introduction to Economic Consequences

Rising sea levels pose significant economic threats to coastal cities. From property damage and loss of tourism revenue to the hefty costs of adaptation measures, the financial implications are vast and varied. This chapter delves into these economic consequences, providing real-life examples from around the world to illustrate the stakes and the need for urgent action.

Property Damage

One of the most immediate and visible economic impacts of rising sea levels is property damage. Coastal properties, whether residential, commercial, or industrial, are at risk from increased flooding, storm surges, and erosion.

Residential Impact:

Coastal homeowners face the loss of property value as their homes become more susceptible to flooding. In Miami, Florida, for example, properties in low-lying areas have seen a decline in value as buyers become wary of future sea level rise. Insurance premiums have skyrocketed, and in some cases, homeowners find it difficult to obtain coverage at all.

Commercial and Industrial Impact:

Businesses located in coastal areas are also vulnerable. The Port of Los Angeles, one of the busiest ports in the United States, faces the threat of rising sea levels, which could disrupt operations and lead to significant economic losses. The cost of retrofitting infrastructure and implementing flood defenses is substantial but necessary to ensure continued operations.

Infrastructure Damage:

Critical infrastructure such as roads, bridges, and public utilities are at risk. In New York City, the aftermath of Hurricane Sandy in 2012 demonstrated the vulnerability of the city's infrastructure to extreme weather events exacerbated by rising sea levels. The cost of repairing and upgrading this infrastructure runs into billions of dollars.

Loss of Tourism Revenue

Tourism is a major economic driver for many coastal cities. The allure of beaches, marine life, and waterfront attractions draws millions of visitors annually. However, rising sea levels threaten to erode these natural assets and diminish their appeal.

Beach Erosion:

In places like the Maldives, a popular tourist destination known for its pristine beaches and clear waters, beach erosion is a significant concern. As beaches disappear, so does the primary attraction for tourists, leading to a decline in visitor numbers and revenue.

Coral Reef Degradation:

Coral reefs, which are crucial for marine biodiversity and a major draw for tourists, are being damaged by rising sea temperatures and acidification. The Great Barrier Reef in Australia has experienced widespread coral bleaching, impacting tourism as fewer healthy reefs remain for diving and snorkeling.

Cultural Heritage Sites:

Coastal cities rich in cultural heritage face the risk of losing historical sites to rising waters. Venice, Italy, known for its historic architecture and canals, is grappling with frequent flooding. Efforts to protect and preserve these sites are costly, and the potential loss of tourism revenue if these sites are damaged is immense.

Costs of Adaptation Measures

Adapting to rising sea levels requires substantial investment in infrastructure and planning. These costs, while necessary, place a significant financial burden on governments and communities.

Sea Walls and Barriers:

Building and maintaining sea walls and storm surge barriers is one of the primary methods of protecting coastal cities. The Netherlands' Delta Works project, a series of dams and barriers designed to protect against flooding, is one of the most extensive and expensive examples, costing billions of euros.

Floodproofing Buildings:

Retrofitting buildings to withstand flooding involves significant costs. In New York City, post-Hurricane Sandy, efforts to floodproof critical infrastructure and public housing have required large-scale investments.

Relocation Costs:

In some cases, retreating from vulnerable coastal areas is the most viable option. This involves relocating entire communities, which can be logistically complex and financially draining. The village of Newtok in Alaska is undergoing such a relocation due to severe erosion and flooding, with costs projected to exceed $100 million.

Real-Life Examples

Several cities around the world are already grappling with the economic impacts of rising sea levels, providing valuable lessons and case studies.

Miami, Florida:

Miami is often cited as one of the most vulnerable cities to sea level rise in the United States. The city's real estate market is directly affected, with properties in flood-prone areas losing value. The local government has committed to significant investment in flood defenses and drainage systems, but the costs are immense.

Bangkok, Thailand:

Bangkok, built on a low-lying delta, faces severe flooding risks exacerbated by rising sea levels. The city has invested in flood barriers and improved drainage systems, but the costs continue to rise as the threats increase. The economic impact is felt across various sectors, from real estate to tourism.

Kiribati:

The Pacific island nation of Kiribati is on the frontlines of sea level rise. The government has purchased land in Fiji as a potential relocation site for its citizens, acknowledging the possibility that their homeland may become uninhabitable. The financial and logistical challenges of such a move are staggering.

The Imperative of Economic Planning

The chapter concludes by emphasizing the importance of proactive economic planning and investment in adaptation measures. The cost of inaction far outweighs the investments required to protect coastal communities. Governments, businesses, and individuals must work together to develop and implement strategies that mitigate the economic impacts of rising sea levels.

In "The Economic Impact," the narrative underscores the financial stakes involved in addressing rising sea levels. Through real-life examples and detailed analysis, the chapter highlights the urgent need for comprehensive economic planning and investment to safeguard the future of coastal cities. The discussion sets the stage for exploring the social and human costs of rising sea levels in the subsequent chapters, reinforcing the multifaceted nature of this global challenge.

Chapter 5
The Human Cost

Introduction to Human Impact

While the economic consequences of rising sea levels are significant, the human cost is equally profound. This chapter delves into the personal stories of individuals and families affected by rising sea levels, bringing a human face to the crisis. From displacement and the loss of livelihoods to the emotional and social impacts, these stories highlight the real-world consequences of this environmental challenge.

Displacement and Relocation

Displacement due to rising sea levels is one of the most devastating human impacts. Entire communities are forced to leave their homes, losing not only their physical space but also their cultural and social ties.

The Village of Newtok, Alaska:

Newtok is a small village on the Ninglick River, home to the indigenous Yup'ik people. Due to severe erosion and permafrost melt caused by rising temperatures, the village is sinking and flooding regularly. The community has decided to relocate to a new site called Mertarvik, several miles away. The relocation process is fraught with challenges, including securing funding, building new infrastructure, and maintaining community cohesion. Elder Lucy Anne, who has lived in Newtok all her life, shares her bittersweet feelings about leaving her ancestral home: "Our roots are here, our memories are here. Moving is like leaving a part of ourselves behind."

The Sinking Islands of Kiribati:

The island nation of Kiribati in the Pacific Ocean is one of the most vulnerable countries to sea level rise. With much of its land only a few meters above sea level, frequent flooding and saltwater intrusion are making life increasingly difficult. Many families are preparing for the possibility of becoming climate refugees. Anote Tong, a former president of Kiribati, has been a vocal advocate for his people: "Our identity is tied to our land. Losing our islands means losing our culture and heritage. We are not just fighting for our homes, but for our existence as a nation."

Loss of Livelihoods

Rising sea levels threaten the livelihoods of millions of people who depend on coastal and marine resources.

Fishermen in the Philippines:

The coastal communities in the Philippines, such as those in the province of Leyte, heavily rely on fishing for their livelihood. Rising sea levels and changing ocean conditions have led to declining fish stocks and more difficult fishing conditions. Mario, a fisherman from Tacloban, describes the struggle: "Every year, it gets harder to catch enough fish to feed my family. The sea is changing, and we are losing our way of life. I don't know how much longer we can continue like this."

Farmers in Bangladesh:

In Bangladesh, rising sea levels and increased salinity in the soil are rendering agricultural lands unproductive. This has devastating effects on farmers who depend on these lands for their income and food supply. Shahana, a farmer from the Satkhira district, speaks about the challenges: "Our crops are failing, and we have no other means to support ourselves. Many families are being forced to move to the cities, where life is even harder. We feel helpless."

Emotional and Social Impacts

Beyond the tangible losses, the emotional and social impacts of rising sea levels are profound. The stress of displacement, loss of livelihood, and uncertainty about the future takes a toll on mental health and community well-being.

Psychological Stress:

The constant threat of flooding and the fear of losing one's home can lead to chronic stress and anxiety. In Miami, Florida, residents of flood-prone neighborhoods experience this stress regularly. Maria, a single mother living in Little Haiti, shares her experience: "Every time it rains heavily, I worry about flooding. I can't sleep, thinking about what would happen if we had to evacuate. It's a constant fear that hangs over us."

Social Fragmentation:

Displacement can lead to the fragmentation of communities, as people are forced to move to different locations. This loss of social networks can exacerbate feelings of isolation and loss. In the Mekong Delta of Vietnam, where rising sea levels threaten entire villages, families are being split apart as younger members move to cities in search of work. Elder Nguyen speaks about the impact on his community: "Our village is like a big family. But now, many young people are leaving, and we are losing the bonds that hold us together. It's heartbreaking."

Voices of Resilience

Despite the immense challenges, there are stories of resilience and hope. Communities are finding ways to adapt and support each other in the face of rising sea levels.

Community Initiatives in Bangladesh:

In Bangladesh, organizations like BRAC are helping coastal communities adapt through initiatives like floating gardens, which allow farmers to grow crops on water. Salma, a participant in the program, shares her gratitude: "These floating gardens have given us a new way to survive. We are learning to adapt and support each other, and it gives us hope for the future."

Youth Activism in the Pacific Islands:

Young activists in the Pacific Islands are raising their voices to demand global action on climate change. Brianna Fruean, a climate activist from Samoa, is part of the Pacific Climate Warriors, a group advocating for the rights and future of island nations. She speaks passionately about their mission: "We are not drowning, we are fighting. Our generation is standing up to protect our homes, our cultures, and our futures. We will not be silent."

Conclusion: The Urgency of Human-Centered Solutions

The chapter concludes by emphasizing the need for human-centered solutions to address the impacts of rising sea levels. Policies and adaptation measures must consider the social, emotional, and cultural dimensions of the crisis. Supporting displaced communities, safeguarding livelihoods, and promoting mental health are essential components of a comprehensive response.

In "The Human Cost," personal stories bring the reality of rising sea levels to life, highlighting the profound impact on individuals and communities. By focusing on the human dimension, the chapter underscores the urgency of addressing this crisis with empathy and action. The narrative sets the stage for exploring the role of governments, technological innovations, and global cooperation in subsequent chapters, reinforcing the multifaceted nature of the challenge.

Chapter 6
The Role of Governments

Introduction to Governmental Response

The escalating threat of rising sea levels necessitates a robust and coordinated response from governments around the world. This chapter scrutinizes government policies and international agreements aimed at combating climate change and rising sea levels. By examining the successes and failures of various initiatives, the political challenges involved, and the potential for future action, this chapter provides a comprehensive understanding of the governmental role in addressing this global crisis.

National Policies and Initiatives

National governments play a critical role in implementing policies and measures to mitigate the impacts of rising sea levels. Different countries have adopted various strategies, reflecting their unique vulnerabilities and capacities.

The Netherlands:

Known for its pioneering flood management systems, the Netherlands has implemented extensive measures to protect its low-lying lands from the sea. The Delta Works, a series of dams, sluices, locks, dikes, and storm surge barriers, is one of the most sophisticated flood defense systems in the world. The Dutch government continuously invests in research, infrastructure, and innovative technologies to stay ahead of rising sea levels. Despite the high costs, the benefits in terms of safety and economic stability are immeasurable.

Bangladesh:

Bangladesh faces significant threats from rising sea levels due to its low elevation and dense population. The government has undertaken various initiatives to enhance resilience, including the construction of cyclone shelters, embankments, and early warning systems. Additionally, projects like floating agriculture and salt-tolerant crops aim to support livelihoods affected by increased salinity. While these efforts have made progress, challenges remain in terms of funding, implementation, and reaching all vulnerable communities.

United States:

The U.S. approach to addressing rising sea levels varies widely across states and municipalities. Coastal cities like New York and Miami have implemented substantial adaptation measures, such as seawalls, elevated infrastructure, and improved drainage systems. Federal initiatives, such as the National Flood Insurance Program (NFIP) and grants from agencies like FEMA, support local efforts. However, political divisions and varying levels of commitment across different administrations and states pose significant challenges to a cohesive national strategy.

International Agreements and Cooperation

International cooperation is crucial for addressing the global nature of climate change and rising sea levels.

Various international agreements and organizations facilitate coordinated action among nations.

Paris Agreement:

Adopted in 2015, the Paris Agreement is a landmark international treaty aimed at limiting global warming to well below 2 degrees Celsius above pre-industrial levels. By committing to nationally determined contributions (NDCs), countries pledge to reduce greenhouse gas emissions and enhance climate resilience. While the agreement represents significant progress, its success depends on the commitment and follow-through of individual countries. Critics argue that current NDCs are insufficient to meet the agreement's goals, highlighting the need for more ambitious action.

United Nations Framework Convention on Climate Change (UNFCCC):

The UNFCCC provides a platform for international dialogue and cooperation on climate change. Its annual Conferences of the Parties (COP) bring together governments, NGOs, and other stakeholders to negotiate and review progress. Initiatives like the Green Climate Fund, established under the UNFCCC, aim to support developing countries in their adaptation and mitigation efforts. However, political and financial challenges often hinder the effectiveness of these initiatives.

Small Island Developing States (SIDS) Coalition:

SIDS are among the most vulnerable to rising sea levels and have formed coalitions to advocate for their interests in international forums. These countries emphasize the need for urgent global action, financial support, and technology transfer to address their unique challenges. The Alliance of Small Island States (AOSIS) has been instrumental in raising awareness and pushing for stronger climate commitments.

Political Challenges and Barriers

Addressing rising sea levels through governmental action is fraught with political challenges and barriers.

Economic Interests vs. Environmental Protection:

Balancing economic development and environmental protection is a perennial challenge. Industries such as fossil fuels, real estate, and tourism often have significant influence over political decisions. Policies that threaten these industries' interests can face strong opposition. For example, efforts to reduce greenhouse gas emissions may be resisted by the oil and gas sector, leading to political gridlock.

Short-Term vs. Long-Term Planning:

Political cycles and the focus on short-term gains often impede long-term planning necessary for addressing climate change. Elected officials may prioritize immediate economic and political benefits over long-term environmental sustainability. This short-sightedness can delay the implementation of crucial adaptation and mitigation measures.

International Cooperation and Equity:

Ensuring equity in international cooperation is a major challenge. Developing countries, which are often the most affected by climate change, have limited resources to implement adaptation measures. Developed countries have a historical responsibility for higher emissions and are expected to provide financial and technological support. However, disparities in contributions and commitments can lead to tensions and hinder progress.

Case Studies of Success and Failure

Examining specific case studies of governmental action provides insights into what works and what doesn't.

Success: Rotterdam, Netherlands:

Rotterdam is a prime example of successful urban adaptation to rising sea levels. The city has implemented a comprehensive strategy that includes flood barriers, water plazas, green roofs, and adaptive architecture. These measures not only enhance resilience but also improve urban living conditions. Rotterdam's approach demonstrates the effectiveness of integrated, multi-faceted solutions and proactive planning.

Failure: Jakarta, Indonesia:

Jakarta faces severe flooding and sinking due to a combination of rising sea levels and excessive groundwater extraction. Despite various initiatives, including the construction of sea walls and planned relocation of the capital, the city's response has been hampered by inadequate funding, corruption, and lack of coordination.

Jakarta's struggles highlight the importance of good governance, adequate resources, and comprehensive planning.

The Path Forward

The chapter concludes by emphasizing the need for strengthened governmental action and international cooperation. Key recommendations for moving forward include:

Enhanced Commitments:

Countries must enhance their commitments to reducing greenhouse gas emissions and investing in adaptation measures. This requires political will, public support, and international pressure.

Sustainable Financing:

Innovative financing mechanisms, such as climate bonds, carbon pricing, and international funds, are essential to support adaptation and mitigation efforts, especially in vulnerable regions.

Integrated Approaches:

Governments should adopt integrated, multi-faceted approaches that combine infrastructure development, environmental conservation, and community engagement. Collaboration across sectors and levels of government is crucial.

Equity and Justice:

Ensuring that vulnerable communities and countries receive adequate support is a moral and practical imperative. This includes financial aid, technology transfer, and capacity building to enhance resilience and adaptive capacity.

In "The Role of Governments," the narrative explores the critical role that governmental policies and international agreements play in addressing rising sea levels. By examining successes, failures, and the political challenges involved, the chapter underscores the necessity of robust and coordinated action to safeguard the future of coastal communities worldwide. The discussion sets the stage for exploring technological innovations, community-based solutions, and global cooperation in the subsequent chapters, highlighting the multifaceted nature of the challenge.

Chapter 7
Technological Innovations

Introduction to Technological Solutions

As the threat of rising sea levels looms larger, innovative technological solutions are crucial for mitigating its effects. This chapter explores various cutting-edge technologies and approaches that offer hope for adaptation and resilience. From robust sea walls and floating cities to advanced warning systems and sustainable urban planning, these innovations represent the forefront of humanity's efforts to cope with the challenges posed by rising sea levels.

Sea Walls and Coastal Barriers

One of the most direct methods of protecting coastal areas from rising sea levels and storm surges is the construction of sea walls and coastal barriers. These structures are designed to prevent flooding and erosion, safeguarding communities and infrastructure.

The Thames Barrier, London:

The Thames Barrier is one of the largest movable flood barriers in the world, protecting London from tidal surges. Completed in 1982, it consists of ten steel gates that can be raised to block the river during high tides or storm surges. The barrier has been raised over 180 times since its completion, demonstrating its effectiveness in protecting one of the world's major cities.

Maeslantkering, Netherlands:

Part of the Delta Works project, the Maeslantkering is a massive storm surge barrier near Rotterdam. It is a movable barrier that closes during extreme weather events, protecting the highly populated and economically significant area from flooding. This advanced engineering marvel showcases the Netherlands' commitment to using technology to defend against rising sea levels.

Seawall in Galveston, Texas:

After the devastating hurricane of 1900, Galveston constructed a 10-mile-long seawall to protect the city from future storm surges. Over the years, it has been extended and reinforced, proving its value in safeguarding the city from numerous hurricanes and high tides.

Floating Cities and Adaptive Architecture

As traditional land-based solutions may become insufficient, particularly in low-lying and densely populated areas, innovative architectural designs and concepts like floating cities are gaining traction.

The Floating City Project, French Polynesia:

In response to rising sea levels threatening island nations, the Seasteading Institute proposed the Floating City Project. This initiative aims to create self-sustaining floating communities that can rise and fall with the sea levels. The project envisions modular, eco-friendly structures that provide housing, commercial spaces, and green areas.

Floating Pavilions, Rotterdam:

Rotterdam, known for its innovative approaches to water management, is home to floating pavilions that serve as multi-functional spaces. These structures are designed to adapt to changing water levels and showcase sustainable living practices. They are energy-efficient, using solar power and rainwater harvesting, and serve as a model for future floating architecture.

Amphibious Houses, Maasbommel, Netherlands:

These houses are built on floating foundations that allow them to rise and fall with water levels. In normal conditions, they rest on the ground, but during floods, they float, tethered to vertical guideposts. This adaptive architecture provides a practical solution for areas prone to flooding, offering safety without the need for permanent relocation.

Advanced Warning Systems

Early warning systems are essential for minimizing the impact of rising sea levels and related extreme weather events. Advances in technology have significantly improved the accuracy and timeliness of these warnings.

The Global Sea Level Observing System (GLOSS):

GLOSS, coordinated by the Intergovernmental Oceanographic Commission (IOC) of UNESCO, is a global network of tide gauges that monitor sea levels. This system provides critical data for predicting sea level rise and informing coastal management decisions.

NOAA's Early Warning System:

The National Oceanic and Atmospheric Administration (NOAA) in the United States operates a sophisticated early warning system that uses satellite data, tide gauges, and computer models to predict storm surges and flooding. This system provides real-time information to emergency management agencies and the public, enabling timely evacuations and preparations.

Tsunami Warning Systems:

Tsunamis, often triggered by underwater earthquakes, pose a significant threat to coastal communities. Advanced tsunami warning systems, such as those operated by the Pacific Tsunami Warning Center (PTWC), use seismic data and ocean buoys to detect and predict tsunamis. These systems provide crucial lead time for evacuations, potentially saving thousands of lives.

Sustainable Urban Planning

Long-term resilience to rising sea levels requires a holistic approach to urban planning that incorporates sustainability, adaptability, and community engagement.

The Sponge City Concept, China:

China's Sponge City initiative aims to improve urban water management through green infrastructure that absorbs, stores, and purifies rainwater. Cities like Wuhan and Shenzhen have implemented measures such as permeable pavements, green roofs, and constructed wetlands. These features reduce flooding, enhance water quality, and provide green spaces for residents.

Green Infrastructure in Copenhagen:

Copenhagen has invested in green infrastructure to manage stormwater and reduce flooding. Projects like the Cloudburst Management Plan include green streets, parks designed to act as reservoirs during heavy rainfall, and underground water storage. These measures not only protect the city from flooding but also enhance urban livability and biodiversity.

Resilient Reefs Initiative:

Coral reefs provide natural protection against storm surges and erosion. The Resilient Reefs Initiative focuses on enhancing the resilience of coral reefs through conservation and restoration efforts. By protecting these natural barriers, coastal communities can reduce their vulnerability to rising sea levels.

The Role of Technology in Community Engagement

Technology also plays a crucial role in engaging communities and raising awareness about the risks and adaptation strategies related to rising sea levels.

Interactive Maps and Tools:

Online platforms like Climate Central's Surging Seas and NOAA's Sea Level Rise Viewer allow individuals to visualize the potential impact of rising sea levels on their communities. These tools empower residents to understand their risks and advocate for necessary measures.

Citizen Science Initiatives:

Programs like the Community Collaborative Rain, Hail and Snow Network (CoCoRaHS) involve citizens in collecting weather data, enhancing the understanding of local climate patterns. Engaging communities in scientific data collection fosters a sense of ownership and awareness about environmental changes.

Virtual Reality (VR) Simulations:

VR technology is being used to create immersive experiences that help people understand the potential impacts of rising sea levels. Projects like the Stanford Virtual Human Interaction Lab's "Our Ocean" allow users to experience sea level rise scenarios, fostering empathy and motivating action.

Conclusion: The Promise of Innovation

The chapter concludes by emphasizing the promise of technological innovation in addressing the challenges posed by rising sea levels. While no single solution can fully mitigate the impacts, the combination of advanced engineering, adaptive architecture, early warning systems, and sustainable urban planning offers a multifaceted approach to building resilience.

In "Technological Innovations," the narrative explores the cutting-edge solutions that are paving the way for a more resilient future. By showcasing successful examples and highlighting the potential of emerging technologies, the chapter underscores the critical role of innovation in safeguarding coastal communities. This discussion sets the stage for examining community-based solutions, global cooperation, and the need for a collective response in the subsequent chapters, highlighting the interconnected nature of the challenge.

Chapter 8
Coastal Ecosystems at Risk

Introduction to Coastal Ecosystems

Coastal ecosystems, including coral reefs, mangroves, and wetlands, provide essential services that support biodiversity, protect shorelines, and sustain human communities. However, rising sea levels pose unprecedented threats to these vital ecosystems. This chapter examines the impact of rising sea levels on marine and coastal ecosystems and explores how their loss could exacerbate the effects on human communities.

Coral Reefs: The Guardians of the Sea

Coral reefs are among the most diverse and valuable ecosystems on Earth. They provide habitat for a wide variety of marine species, protect coastlines from storm surges, and support fishing and tourism industries.

Impact of Rising Sea Levels:

Rising sea levels can drown coral reefs by increasing water depths beyond the optimal range for coral growth. Additionally, increased sea surface temperatures and ocean acidification, both linked to climate change, exacerbate the stress on coral reefs. Coral bleaching, a phenomenon where corals expel the symbiotic algae living in their tissues due to stress, is becoming more frequent and severe, leading to widespread reef degradation.

Case Study: The Great Barrier Reef:

The Great Barrier Reef, the world's largest coral reef system, has experienced significant bleaching events in recent years. Rising sea temperatures and changing sea levels threaten the delicate balance of this ecosystem. Scientists estimate that nearly half of the reef's coral cover has been lost over the past three decades, with dire implications for the marine species that depend on it and the tourism industry that thrives on its beauty.

Mangroves: Nature's Coastal Defenders

Mangroves are salt-tolerant trees and shrubs that grow in coastal intertidal zones. They play a crucial role in protecting shorelines from erosion, providing habitat for marine life, and sequestering carbon.

Impact of Rising Sea Levels:

Mangroves are particularly vulnerable to rising sea levels because they rely on a delicate balance between freshwater and saltwater. Increased sea levels can lead to the inundation of mangrove forests, altering salinity levels and waterlogging roots. This can result in the die-off of mangrove trees and the loss of critical habitat for numerous species.

Case Study: Sundarbans Mangrove Forest:

The Sundarbans, the largest mangrove forest in the world, located in the delta region of India and Bangladesh, is facing severe threats from rising sea levels. The encroaching seawater and increased salinity are killing off mangrove trees, reducing the forest's ability to protect coastal communities from storm surges and erosion. The loss of the Sundarbans would not only impact biodiversity but also jeopardize the livelihoods of millions of people who depend on its resources.

Wetlands: The Natural Water Purifiers

Wetlands, including marshes, swamps, and bogs, are critical ecosystems that provide water filtration, flood control, and habitat for wildlife. They act as natural buffers against storm surges and rising waters.

Impact of Rising Sea Levels:

Wetlands are highly susceptible to rising sea levels because they exist at the interface of land and water. Increased sea levels can lead to the inundation and loss of wetlands, reducing their capacity to absorb floodwaters and filter pollutants. This can result in increased flooding, water pollution, and loss of biodiversity.

Case Study: Louisiana Wetlands:

The wetlands of Louisiana, particularly in the Mississippi River Delta, are disappearing at an alarming rate due to a combination of rising sea levels, subsidence, and human activities such as oil and gas extraction. These wetlands provide critical protection against hurricanes and support rich biodiversity. The loss of Louisiana's wetlands not only increases the vulnerability of coastal communities to storms but also impacts fisheries and water quality.

The Interconnectedness of Ecosystems and Human Communities

The health of coastal ecosystems is intrinsically linked to the well-being of human communities. The loss of coral reefs, mangroves, and wetlands can exacerbate the effects of rising sea levels on people in several ways:

Increased Flooding and Erosion:

Coastal ecosystems act as natural barriers against storm surges and erosion. Their degradation can lead to more frequent and severe flooding, putting lives and property at risk. Without the protection of these ecosystems, coastal communities become more vulnerable to the impacts of extreme weather events.

Loss of Livelihoods:

Many communities depend on coastal ecosystems for their livelihoods. Fishing, tourism, and agriculture are all sectors that benefit from healthy reefs, mangroves, and wetlands. The loss of these ecosystems can lead to economic decline and increased poverty in affected regions.

Decreased Biodiversity:

Coastal ecosystems support a rich diversity of species. Their degradation can lead to the loss of habitat and a decline in biodiversity, impacting ecosystem services such as food provision, water purification, and climate regulation. The loss of biodiversity also reduces the resilience of ecosystems to environmental changes.

Conservation and Restoration Efforts

Efforts to conserve and restore coastal ecosystems are crucial for mitigating the impacts of rising sea levels and preserving their benefits for future generations.

Coral Reef Restoration:

Initiatives like coral gardening and artificial reefs aim to restore damaged coral ecosystems. Coral gardening involves growing corals in nurseries and transplanting them to degraded reefs. Artificial reefs, made from materials like concrete and steel, provide structures for coral larvae to settle and grow. These efforts help rebuild reef structures and support marine life.

Mangrove Reforestation:

Planting mangroves in areas where they have been lost can help restore their protective functions. Mangrove reforestation projects, such as those in the Philippines and Indonesia, involve community participation and education to ensure the success and sustainability of these efforts. Healthy mangrove forests can stabilize shorelines, support fisheries, and sequester carbon.

Wetland Restoration:

Wetland restoration projects focus on re-establishing natural hydrology, planting native vegetation, and removing invasive species. Programs like the Everglades Restoration in Florida aim to restore the natural flow of water and enhance the resilience of wetland ecosystems. Restored wetlands can provide flood protection, improve water quality, and support wildlife.

The Role of Policy and Community Engagement

Policy measures and community engagement are essential for the effective conservation and restoration of coastal ecosystems.

Protected Areas:

Establishing marine protected areas (MPAs) can safeguard critical habitats from destructive activities. MPAs provide refuges for marine life, allowing ecosystems to recover and thrive. Policies that enforce sustainable fishing practices and restrict coastal development are crucial for maintaining the health of these ecosystems.

Community Involvement:

Engaging local communities in conservation efforts ensures their success and sustainability. Community-based management approaches, such as those in Fiji and Belize, involve local residents in monitoring and protecting coastal ecosystems. Education and awareness programs empower communities to take an active role in conservation.

International Cooperation:

Global initiatives like the Ramsar Convention on Wetlands and the International Coral Reef Initiative (ICRI) promote the conservation and sustainable use of coastal ecosystems. International cooperation and funding support the implementation of conservation projects in developing countries, where resources may be limited.

Conclusion: Preserving Nature's Defenses

The chapter concludes by emphasizing the importance of preserving and restoring coastal ecosystems as natural defenses against rising sea levels. The health of coral reefs, mangroves, and wetlands is critical for the resilience of both ecosystems and human communities.

In "Coastal Ecosystems at Risk," the narrative explores the profound impact of rising sea levels on vital coastal ecosystems and highlights the interconnectedness of these ecosystems with human well-being. By examining the threats, conservation efforts, and the role of policy and community engagement, the chapter underscores the urgency of protecting these natural defenses. The discussion sets the stage for exploring community-based solutions, global cooperation, and the need for a collective response in subsequent chapters, highlighting the multifaceted nature of the challenge.

Chapter 9
Voices from the Frontlines

Introduction to Grassroots Efforts

Amidst the global threat of rising sea levels, countless individuals are on the frontlines, striving to protect their communities. This chapter features interviews with activists, scientists, and community leaders who are spearheading grassroots efforts and implementing local resilience strategies. Their stories and insights highlight the importance of community engagement and the innovative solutions being employed to combat this crisis.

Activists: Champions of Change

Activists play a crucial role in raising awareness and mobilizing communities to take action against the impacts of rising sea levels. They advocate for sustainable practices, policy changes, and community-based adaptation measures.

1. Greta Thunberg – Global Climate Activist

Greta Thunberg, the Swedish environmental activist, has become a global symbol of the youth-led climate movement. Her activism emphasizes the urgent need for action to mitigate climate change, which directly impacts sea levels. In an interview, she shares her perspective on how individual actions and grassroots movements can drive significant change:

"Every action counts. It's not just about the big policy changes, but also about what we do in our daily lives. We need to pressure our leaders, but we also need to lead by example. Coastal communities, especially, need to adapt and innovate, showing the world that resilience and sustainability are possible."

2. Atiq Rahman – Bangladeshi Environmentalist

Dr. Atiq Rahman, an environmental scientist and activist from Bangladesh, has dedicated his life to addressing the impacts of climate change on vulnerable communities. He discusses the unique challenges faced by Bangladesh, a country highly susceptible to rising sea levels:

"In Bangladesh, we are living the reality of climate change. Our coastal areas are facing constant threats from sea level rise and storm surges. But we are not passive victims. Through community-based adaptation projects, such as floating farms and homes, we are finding ways to live with the changes. Our resilience is rooted in innovation and community solidarity."

Scientists: Pioneers of Knowledge

Scientists are at the forefront of understanding the mechanisms of rising sea levels and developing technologies and strategies to mitigate their impacts. Their research informs policies and practical solutions that enhance community resilience.

3. Dr. Katharine Hayhoe – Climate Scientist

Dr. Katharine Hayhoe, a prominent climate scientist and communicator, discusses the scientific understanding of sea level rise and its implications for coastal cities. Her work bridges the gap between science and public awareness:

"The science is clear: sea levels are rising at an accelerating rate due to the melting of ice caps and the thermal expansion of seawater. Coastal cities must prepare for these changes through both mitigation and adaptation. This means reducing emissions to slow the rate of rise and implementing measures such as building sea walls, restoring wetlands, and planning for managed retreats where necessary."

4. Dr. Benjamin Horton – Coastal Geologist

Dr. Benjamin Horton, an expert in sea level rise and coastal hazards, shares his insights on the importance of integrating scientific research with local knowledge. He highlights successful projects that combine scientific expertise with community engagement:

"One of our key projects involves working with indigenous communities in the Pacific Islands. These communities possess invaluable traditional knowledge about living with the sea. By combining their insights with our scientific data, we develop tailored adaptation strategies that are both effective and culturally appropriate. This collaborative approach is essential for building resilience."

Community Leaders: Builders of Resilience

Community leaders are instrumental in implementing local adaptation strategies and fostering a sense of unity and purpose. Their efforts ensure that solutions are not only effective but also inclusive and sustainable.

5. Eulogia González – Leader in Puerto Rico

Eulogia González, a community leader in Puerto Rico, has been at the forefront of recovery and resilience efforts following the devastation of Hurricane Maria. She discusses the importance of community-driven initiatives:

"After Hurricane Maria, we realized that we couldn't wait for outside help. We had to take matters into our own hands. We organized community gardens to ensure food security, built rainwater harvesting systems, and created local emergency response teams. By empowering our community, we are better prepared for future challenges. Resilience is about self-reliance and mutual support."

6. Chief Alvino Fantini – Leader of the Krenak People in Brazil

Chief Alvino Fantini, a leader of the Krenak indigenous people in Brazil, discusses how his community is adapting to the impacts of climate change while preserving their cultural heritage. He emphasizes the importance of traditional knowledge in resilience strategies:

"Our people have always lived in harmony with nature. The rising waters are a challenge, but they are not new to us. We use traditional knowledge to adapt our agriculture, manage water resources, and protect our sacred lands. It's essential that we combine our ancestral wisdom with modern techniques to build a sustainable future for our community."

Case Studies of Grassroots Efforts

7. The Maldives: Building Artificial Islands

The Maldives, an island nation highly vulnerable to rising sea levels, has undertaken ambitious projects to ensure its survival. One such initiative involves building artificial islands that are elevated above the natural sea level. These islands serve as safe havens for residents and critical infrastructure. A local engineer involved in the project shares:

"We are literally building our future. By creating these artificial islands, we are providing a long-term solution to the threats posed by rising sea levels. It's a combination of engineering and community planning that ensures we can continue to live in our beautiful homeland."

8. Louisiana: Community-Led Wetland Restoration

In Louisiana, community groups are leading efforts to restore wetlands that act as natural barriers against flooding. One such initiative, led by the Barataria-Terrebonne National Estuary Program, focuses on planting native vegetation and building oyster reefs to stabilize the shoreline. A program coordinator explains:

"Restoring our wetlands is a community effort. Volunteers from all walks of life come together to plant marsh grasses, build oyster reefs, and monitor the health of our wetlands. These efforts not only protect our homes from flooding but also enhance the health of our ecosystem. It's a testament to what we can achieve when we work together."

The Power of Collective Action

The chapter concludes by highlighting the power of collective action and the importance of local resilience strategies in addressing the challenges posed by rising sea levels. The stories and insights from activists, scientists, and community leaders demonstrate that while the threat is global, the solutions are often local and rooted in the strength and ingenuity of communities.

In "Voices from the Frontlines," the narrative brings to life the experiences and perspectives of those directly engaged in the fight against rising sea levels. By showcasing grassroots efforts and local resilience strategies, the chapter underscores the critical role of community engagement and the innovative solutions being employed. This discussion sets the stage for exploring global cooperation and the need for a collective response in subsequent chapters, highlighting the interconnected nature of the challenge.

Chapter 10
The Global Perspective

Introduction to the Global Threat

Rising sea levels are a universal challenge, affecting countries across the globe regardless of their geographical location, economic status, or development level. This chapter delves into how different countries are affected by rising sea levels and showcases diverse strategies employed to combat this threat. Through case studies from Asia, Africa, Europe, and the Americas, we gain a comprehensive understanding of the global nature of this issue and the innovative approaches being adopted worldwide.

Asia: The Vulnerable Giant

Asia, home to some of the world's most populous and economically significant regions, faces severe threats from rising sea levels. Coastal cities and island nations are particularly vulnerable.

1. Bangladesh: Living on the Edge

Impact:

Bangladesh is one of the countries most affected by rising sea levels due to its low-lying geography and high population density. Frequent flooding, coastal erosion, and saltwater intrusion threaten agriculture, drinking water, and homes.

Response:

The country has implemented several adaptation strategies, including building embankments, developing salt-tolerant crops, and creating floating farms and schools. Community-based adaptation projects, supported by international organizations, focus on enhancing local resilience.

"Our survival depends on our ability to adapt. We are innovating with floating farms and homes, and engaging communities in resilience-building activities. International cooperation is crucial for providing the resources and knowledge we need." - Dr. Saleemul Huq, climate scientist and director of the International Centre for Climate Change and Development (ICCCAD).

2. Japan: Engineering Solutions

Impact:

Japan, an island nation with extensive coastal cities, faces threats from both rising sea levels and increased frequency of typhoons and storm surges.

Response:

Japan employs advanced engineering solutions such as sea walls, levees, and flood gates. The country has also invested in comprehensive early warning systems and disaster preparedness education.

"Our approach combines cutting-edge technology with traditional knowledge. We are continuously improving our infrastructure and emergency response capabilities to protect our communities from the rising sea." - Dr. Kazuhiko Takeuchi, President of the Institute for Global Environmental Strategies (IGES).

Africa: Battling the Elements

Africa's diverse landscapes and varying levels of development present unique challenges in addressing rising sea levels. Coastal cities and island nations are particularly at risk.

3. Nigeria: Urban Resilience in Lagos

Impact:

Lagos, Nigeria's largest city, faces significant threats from rising sea levels, including frequent flooding, coastal erosion, and saltwater intrusion into freshwater supplies.

Response:

The government is implementing urban resilience projects such as the Eko Atlantic City project, which involves building a new city on reclaimed land protected by sea walls. Additionally, community-led initiatives focus on improving drainage systems and raising public awareness about climate adaptation.

"We are transforming our challenges into opportunities by creating resilient infrastructure and empowering communities to adapt. However, sustainable development and international support are essential for our long-term success." - Babatunde Fashola, former Governor of Lagos State.

4. Seychelles: Conservation and Adaptation

Impact:

The Seychelles, an island nation, is highly vulnerable to sea level rise, which threatens its tourism-dependent economy and biodiversity-rich coastal ecosystems.

Response:

The Seychelles government has implemented marine conservation programs, including the creation of marine protected areas and coral reef restoration projects. The country also focuses on sustainable tourism practices and climate-smart agriculture.

"Our survival depends on the health of our marine ecosystems. We are investing in conservation and sustainable development to protect our islands and livelihoods." - Wavel Ramkalawan, President of Seychelles.

Europe: Fortifying the Coasts

Europe's extensive coastlines and densely populated coastal cities face significant risks from rising sea levels. Countries are employing a mix of engineering, policy, and community-based approaches to address these challenges.

5. The Netherlands: A Legacy of Water Management

Impact:

The Netherlands, with much of its land below sea level, has a long history of battling rising waters. Climate change exacerbates the threat, increasing the risk of flooding.

Response:

The Dutch have developed some of the world's most advanced water management systems, including the Delta Works, a series of dams, sluices, locks, dikes, and storm surge barriers. The Room for the River program involves creating floodplains and water storage areas to manage excess water.

"Our approach integrates engineering, nature, and community planning. We continuously innovate to protect our land and people from rising waters." - Henk Ovink, Special Envoy for International Water Affairs for the Kingdom of the Netherlands.

6. Italy: Preserving Venice

Impact:

Venice, Italy, faces severe flooding due to rising sea levels and subsidence. The city, famous for its canals, is at risk of becoming uninhabitable.

Response:

The MOSE project (Modulo Sperimentale Elettromeccanico) involves a series of movable flood barriers designed to protect Venice from high tides and storm surges. Additionally, efforts are underway to restore and maintain the city's historic architecture and infrastructure.

"Venice is a cultural treasure that we must protect. The MOSE project is a critical component of our strategy, but we must also focus on sustainable tourism and urban planning." - Luigi Brugnaro, Mayor of Venice.

The Americas: Adapting to New Realities

North and South American countries face diverse challenges from rising sea levels, from urban flooding to the displacement of indigenous communities.

7. United States: Adapting Miami

Impact:

Miami, Florida, is one of the most vulnerable cities in the United States to sea level rise, facing frequent flooding, saltwater intrusion, and property damage.

Response:

Miami is investing in a range of adaptation measures, including raising roads, installing pump stations, and implementing stringent building codes. The city is also exploring the potential of living shorelines and mangrove restoration to provide natural protection.

"We are taking a multifaceted approach to protect our city. This includes infrastructure improvements, community engagement, and exploring innovative solutions like living shorelines." - Daniella Levine Cava, Mayor of Miami-Dade County.

8. Brazil: Protecting Coastal Communities

Impact:

Brazil's extensive coastline is home to many vulnerable communities, particularly in low-lying areas where rising sea levels threaten homes, agriculture, and biodiversity.

Response:

Brazil is focusing on restoring mangroves and other coastal ecosystems to provide natural protection against rising waters. Community-based adaptation projects, such as the restoration of traditional fishing techniques, are also being implemented.

"Our coastal communities are resilient and resourceful. By restoring our natural defenses and integrating traditional knowledge with modern practices, we are building a sustainable future." - Marina Silva, Brazilian environmentalist and politician.

Conclusion: A Unified Response

The chapter concludes by emphasizing the need for a unified global response to rising sea levels. While the strategies employed by different countries vary based on their unique circumstances, the underlying principles of resilience, innovation, and community engagement are universal. International cooperation, knowledge sharing, and sustained investment in adaptation measures are critical for addressing this global challenge.

In "The Global Perspective," the narrative explores the diverse impacts and responses to rising sea levels across different regions. By highlighting case studies from Asia, Africa, Europe, and the Americas, the chapter underscores the interconnected nature of this issue and the importance of a collective response. This discussion sets the stage for exploring the role of international organizations, policy frameworks, and future directions in the subsequent chapters, highlighting the multifaceted nature of the challenge.

Chapter 11
Urban Planning for a Watery Future

Introduction to Urban Planning in the Face of Rising Sea Levels

As rising sea levels threaten coastal cities around the world, urban planners and architects are at the forefront of developing innovative strategies to create resilient and sustainable urban environments. This chapter explores how cities can be designed and restructured to withstand the challenges posed by a changing climate. By integrating resilient infrastructure, sustainable building materials, and green spaces, urban planners are paving the way for a future where cities not only survive but thrive amid rising waters.

Resilient Infrastructure: The Backbone of Future Cities

Resilient infrastructure is critical for protecting cities from the impacts of rising sea levels. This involves designing and building structures that can withstand flooding, storm surges, and erosion, while also ensuring the safety and functionality of urban areas.

1. Elevated Buildings and Flood-Resistant Design

Concept:

Elevated buildings and flood-resistant designs are essential for protecting urban areas from rising waters. This includes raising the foundations of new buildings, retrofitting existing structures, and incorporating flood barriers.

Examples:

In New York City, the East Side Coastal Resiliency Project involves elevating parks and constructing floodwalls to protect neighborhoods from storm surges and sea level rise.

In Bangkok, Thailand, architects are designing stilted homes and floating structures to adapt to frequent flooding.

"Elevating buildings and creating flood-resistant designs are crucial steps in safeguarding our cities. By integrating these elements into urban planning, we can ensure that our communities remain safe and functional in the face of rising sea levels." - Dr. Henk Ovink, Special Envoy for International Water Affairs for the Kingdom of the Netherlands.

2. Adaptive Infrastructure: Living with Water

Concept:

Adaptive infrastructure involves designing urban spaces that can coexist with water, rather than simply trying to keep it out. This includes creating floodable parks, water plazas, and permeable surfaces that absorb and manage excess water.

Examples:

In Rotterdam, Netherlands, the Benthemplein Water Plaza serves as both a recreational area and a water storage facility during heavy rainfall.

In Copenhagen, Denmark, the city's Cloudburst Management Plan includes creating green streets and urban spaces that can temporarily store rainwater.

"Adaptive infrastructure allows us to live with water in a harmonious way. By designing cities that can absorb and manage water, we can reduce the risk of flooding and create more resilient urban environments." - Professor Kristina Hill, University of California, Berkeley.

Sustainable Building Materials: Building for the Future

Using sustainable building materials is a key aspect of creating resilient urban environments. These materials not only reduce the environmental impact of construction but also enhance the durability and adaptability of buildings.

3. Green Building Materials

Concept:

Green building materials, such as recycled steel, bamboo, and rammed earth, offer sustainable alternatives to traditional construction materials. These materials are often more durable and have a lower carbon footprint.

Examples:

In Singapore, the Oasia Hotel Downtown is constructed using a combination of green building materials and features a façade covered in plants, which helps manage rainwater and reduce heat.

In Vancouver, Canada, the Brock Commons Tallwood House is one of the tallest mass timber buildings in the world, showcasing the potential of sustainable wood construction.

"Green building materials are essential for creating resilient and sustainable urban environments. By choosing materials that are both durable and environmentally friendly, we can reduce our impact on the planet while building for the future." - Dr. Vivian Loftness, Professor of Architecture at Carnegie Mellon University.

4. Innovative Construction Techniques

Concept:

Innovative construction techniques, such as modular building and 3D printing, allow for more efficient and sustainable construction processes. These techniques can also enhance the adaptability of buildings to changing environmental conditions.

Examples:

In the United Arab Emirates, the Dubai Future Foundation is exploring the use of 3D printing to create affordable and sustainable housing solutions that can be quickly deployed in response to rising sea levels.

In the Netherlands, the Floating Pavilion in Rotterdam is an example of modular construction that can be easily relocated or adapted to changing water levels.

"Innovative construction techniques enable us to build more efficiently and sustainably. By embracing these methods, we can create adaptable structures that are better equipped to handle the challenges of rising sea levels." - Architect Bjarke Ingels, Founder of Bjarke Ingels Group (BIG).

Integrating Green Spaces: Nature as a Buffer

Green spaces play a crucial role in urban planning by providing natural buffers against rising sea levels and enhancing the overall resilience of cities. These spaces also offer social, environmental, and economic benefits.

5. Urban Wetlands and Mangroves

Concept:

Urban wetlands and mangroves act as natural barriers against storm surges and flooding. They absorb excess water, reduce erosion, and provide habitat for wildlife.

Examples:

In New Orleans, the city is investing in wetland restoration projects to protect against hurricanes and rising sea levels. The Bayou Bienvenue Wetland Triangle is one such project that aims to restore the natural wetland ecosystem.

In Jakarta, Indonesia, the city is implementing a mangrove restoration program to combat coastal erosion and flooding.

"Restoring urban wetlands and mangroves is a powerful way to protect our cities from rising sea levels. These natural buffers not only reduce the risk of flooding but also

Conclusion: Preserving Nature's Defenses

The chapter concludes by emphasizing the importance of preserving and restoring coastal ecosystems as natural defenses against rising sea levels. The health of coral reefs, mangroves, and wetlands is critical for the resilience of both ecosystems and human communities.

In "Coastal Ecosystems at Risk," the narrative explores the profound impact of rising sea levels on vital coastal ecosystems and highlights the interconnectedness of these ecosystems with human well-being. By examining the threats, conservation efforts, and the role of policy and community engagement, the chapter underscores the urgency of protecting these natural defenses. The discussion sets the stage for exploring community-based solutions, global cooperation, and the need for a collective response in subsequent chapters, highlighting the multifaceted nature of the challenge.

Chapter 9
Voices from the Frontlines

Introduction to Grassroots Efforts

Amidst the global threat of rising sea levels, countless individuals are on the frontlines, striving to protect their communities. This chapter features interviews with activists, scientists, and community leaders who are spearheading grassroots efforts and implementing local resilience strategies. Their stories and insights highlight the importance of community engagement and the innovative solutions being employed to combat this crisis.

Activists: Champions of Change

Activists play a crucial role in raising awareness and mobilizing communities to take action against the impacts of rising sea levels. They advocate for sustainable practices, policy changes, and community-based adaptation measures.

1. Greta Thunberg – Global Climate Activist

Greta Thunberg, the Swedish environmental activist, has become a global symbol of the youth-led climate movement. Her activism emphasizes the urgent need for action to mitigate climate change, which directly impacts sea levels. In an interview, she shares her perspective on how individual actions and grassroots movements can drive significant change:

enhance biodiversity and provide recreational spaces for residents." - Dr. Robert Twilley, Executive Director of the Louisiana Sea Grant College Program.

6. Green Roofs and Vertical Gardens

Concept:

Green roofs and vertical gardens help manage rainwater, reduce urban heat islands, and improve air quality. These green spaces can be integrated into buildings to enhance their resilience and sustainability.

Examples:

In Chicago, the city has implemented a Green Roof Initiative, encouraging the installation of green roofs on both public and private buildings. The City Hall's rooftop garden is a notable example.

In Milan, Italy, the Bosco Verticale (Vertical Forest) is a pair of residential towers featuring over 900 trees and thousands of plants, which help absorb CO_2, reduce noise pollution, and manage rainwater.

"Green roofs and vertical gardens transform urban environments into greener, more resilient spaces. They offer a range of benefits, from improved air quality to enhanced stormwater management, making our cities more livable and sustainable." - Architect Stefano Boeri, Designer of Bosco Verticale.

Policy and Community Engagement: Driving Change

Effective urban planning for rising sea levels requires supportive policies and active community engagement.

Policies must incentivize sustainable practices, and communities must be involved in planning and decision-making processes.

7. Policy Frameworks for Resilient Cities

Concept:

Governments and municipalities must establish policy frameworks that promote resilient urban planning and sustainable development. This includes zoning laws, building codes, and incentives for green construction.

Examples:

In Singapore, the city-state's Urban Redevelopment Authority has implemented stringent building codes and land-use policies to ensure that new developments are resilient to rising sea levels.

In the United States, the Federal Emergency Management Agency (FEMA) provides guidelines and funding for communities to enhance their resilience to natural disasters, including rising sea levels.

"Policy frameworks are essential for guiding urban development in a way that prioritizes resilience and sustainability. Governments must take a proactive approach to ensure that cities are prepared for the challenges of rising sea levels." - Dr. Michael Beck, Director of the Coastal Resilience Lab at the University of California, Santa Cruz.

8. Community-Led Initiatives

Concept:

Community-led initiatives empower residents to take an active role in planning and implementing resilience strategies. Engaging communities ensures that solutions are tailored to local needs and contexts.

Examples:

In the Philippines, the city of Iloilo has implemented a participatory approach to urban planning, involving residents in the design and execution of flood management projects.

In New York City, the Lower East Side's "Gardens Rising" project combines community gardens with flood mitigation efforts, transforming green spaces into flood-resilient areas.

"Community engagement is crucial for creating resilient urban environments. When residents are involved in planning and decision-making, the solutions are more effective and sustainable." - Dr. Mindy Fullilove, Professor of Urban Policy and Health at The New School.

Conclusion: Building Resilient and Sustainable Cities

The chapter concludes by emphasizing the importance of holistic and integrated approaches to urban planning in the face of rising sea levels. By combining resilient infrastructure, sustainable building materials, green spaces, supportive policies, and community engagement, cities can be designed to withstand the challenges of a watery future.

In "Urban Planning for a Watery Future," the narrative explores the innovative strategies employed by urban planners and architects to create resilient and sustainable cities. By highlighting examples from around the world, the chapter underscores the importance of a multifaceted approach to urban planning. This discussion sets the stage for exploring the role of international cooperation, policy frameworks, and future directions in the subsequent chapters, highlighting the interconnected nature of the challenge.

Chapter 12
The Power of Awareness

Introduction to Awareness and Education

Raising public awareness about the risks of rising sea levels is crucial for fostering a collective response to this global challenge. Effective awareness campaigns and education initiatives can inform people, change behaviors, and inspire action. This chapter explores the role of media, schools, and community organizations in spreading knowledge about rising sea levels and highlights successful efforts from around the world. By understanding the power of awareness, communities can be better prepared to face the realities of a changing climate.

The Role of Media in Raising Awareness

The media plays a pivotal role in informing the public about the dangers of rising sea levels and climate change. Through news reports, documentaries, social media, and other platforms, the media can reach a wide audience and shape public perception.

1. News Coverage and Investigative Reporting

Concept:

News organizations have the ability to bring attention to the impacts of rising sea levels through detailed coverage and investigative reporting. By highlighting real-world examples and expert opinions, they can convey the urgency of the situation.

Examples:

The Guardian's "The Polluted Water We Drink" series investigates the impact of climate change on water quality and availability, including the effects of rising sea levels on coastal communities.

National Geographic's documentaries, such as "Before the Flood," narrated by Leonardo DiCaprio, explore the global effects of climate change and the importance of taking action.

"Through in-depth reporting and compelling storytelling, the media can raise awareness and inspire action on critical issues like rising sea levels. It is our responsibility to inform and engage the public." - Bill McKibben, Environmental Journalist and Founder of 350.org.

2. Social Media Campaigns

Concept:

Social media platforms offer a powerful tool for spreading awareness about rising sea levels. Campaigns on platforms like Twitter, Facebook, and Instagram can reach millions of people, particularly younger audiences.

Examples:

The #ClimateStrike movement, initiated by Greta Thunberg, has mobilized millions of young people around the world to demand action on climate change, including addressing rising sea levels.

The "Years of Living Dangerously" campaign uses social media to share stories and raise awareness about climate impacts, including sea level rise.

"Social media allows us to connect with a global audience and mobilize action on a scale never seen before. By sharing stories and information, we can create a movement for change." - Greta Thunberg, Climate Activist.

Education Initiatives: Building Knowledge and Resilience

Education is a key component in preparing future generations to deal with the impacts of rising sea levels. Schools, universities, and educational programs can equip students with the knowledge and skills needed to address this challenge.

3. Integrating Climate Education in Schools

Concept:

Integrating climate education into school curricula ensures that students learn about the science of climate change, the risks of rising sea levels, and potential solutions from an early age.

Examples:

In Italy, climate change education has been made mandatory in schools, covering topics such as sustainability, environmental science, and the impacts of rising sea levels.

The United Nations Educational, Scientific and Cultural Organization (UNESCO) promotes climate education through its "Education for Sustainable Development" program, which includes resources and guidelines for teaching about climate change.

"By educating students about climate change and its impacts, we empower them to become informed and proactive citizens. Education is the foundation of a resilient and sustainable future." - Audrey Azoulay, Director-General of UNESCO.

4. University Research and Outreach Programs

Concept:

Universities play a crucial role in advancing research on rising sea levels and climate adaptation. Outreach programs can extend this knowledge to communities and policymakers.

Examples:

The University of California, Berkeley's "Climate Readiness Institute" conducts research on climate impacts and engages with local communities to develop adaptation strategies.

The "Sea Level Solutions Center" at Florida International University focuses on studying sea level rise and educating the public through workshops, seminars, and community projects.

"Universities are hubs of knowledge and innovation. Through research and outreach, we can develop solutions and educate the public about the risks of rising sea levels." - Dr. Ben Strauss, CEO and Chief Scientist of Climate Central.

Community Organizations: Grassroots Efforts and Local Initiatives

Community organizations are instrumental in raising awareness and building resilience at the local level. By engaging residents and providing resources, these organizations can drive grassroots efforts to address rising sea levels.

5. Local Advocacy Groups

Concept:

Local advocacy groups work to raise awareness about climate change and rising sea levels within their communities. These groups often organize events, distribute educational materials, and advocate for policy changes.

Examples:

The "Surfrider Foundation" advocates for the protection of oceans and coastal environments, including addressing the impacts of rising sea levels through community engagement and policy advocacy.

"350.org" is a global grassroots movement that organizes campaigns and events to raise awareness about climate change and promote climate action at the local level.

"Grassroots advocacy is crucial for driving change. By mobilizing communities and advocating for policy reforms, we can address the threats posed by rising sea levels." - May Boeve, Executive Director of 350.org.

6. Community Resilience Projects

Concept:

Community resilience projects involve local residents in efforts to adapt to rising sea levels. These projects often focus on building local capacity, enhancing infrastructure, and fostering a sense of community.

Examples:

In Norfolk, Virginia, the "Norfolk Resilience Strategy" engages residents in developing and implementing plans to adapt to sea level rise, including flood management and green infrastructure projects.

The "Resilient Islands" initiative in the Caribbean, led by The Nature Conservancy, works with local communities to develop strategies for managing rising sea levels and protecting coastal ecosystems.

"Community resilience projects empower residents to take action and build a safer, more sustainable future. By working together, we can create communities that are resilient to rising sea levels." - Jennifer Morris, CEO of The Nature Conservancy.

The Power of Art and Culture in Raising Awareness

Art and culture can be powerful tools for raising awareness about rising sea levels and climate change. Through visual arts, literature, music, and other cultural expressions, artists can convey the urgency and emotional impact of the issue.

7. Climate Change Art Installations

Concept:

Art installations and exhibitions focused on climate change can evoke powerful emotional responses and spark conversations about the impacts of rising sea levels.

Examples:

Olafur Eliasson's "Ice Watch" installation involved placing large blocks of melting ice in urban areas to highlight the reality of ice melt and rising sea levels.

"HighWaterLine," a public art project by Eve Mosher, involved drawing a blue chalk line in urban areas to mark future sea level rise, engaging communities in discussions about climate adaptation.

"Art has the ability to move people in ways that data and reports cannot. By creating powerful visual and emotional experiences, we can raise awareness and inspire action on climate change." - Olafur Eliasson, Artist.

8. Climate Change in Literature and Film

Concept:

Literature and film can explore the themes of climate change and rising sea levels, bringing the issue to a wider audience through storytelling and creative expression.

Examples:

Kim Stanley Robinson's novel "New York 2140" envisions a future where rising sea levels have transformed New York

City, blending fiction with climate science to explore the impacts of sea level rise.

The documentary film "An Inconvenient Truth," featuring Al Gore, has been instrumental in raising global awareness about climate change and the urgent need for action.

"Through literature and film, we can tell compelling stories that resonate with people and raise awareness about the profound impacts of rising sea levels." - Kim Stanley Robinson, Author.

Conclusion: The Collective Power of Awareness

The chapter concludes by emphasizing the collective power of awareness in addressing the challenges of rising sea levels. By leveraging the strengths of media, education, community organizations, and cultural expressions, we can inform and inspire people to take action. Awareness is the first step towards building resilient and sustainable communities that can withstand the impacts of a changing climate.

In "The Power of Awareness," the narrative explores how different sectors contribute to raising awareness about rising sea levels. By highlighting successful examples and emphasizing the importance of education and engagement, the chapter underscores the critical role of awareness in fostering a collective response to this global challenge. This discussion sets the stage for exploring the role of policy, technology, and international cooperation in the subsequent chapters, highlighting the multifaceted nature of the challenge.

Chapter 13
Adaptation vs. Mitigation

Introduction: Two Sides of the Climate Change Coin

The challenge of rising sea levels presents a critical question for policymakers, scientists, and communities: Should we focus on adapting to the inevitable changes or mitigating the root causes of climate change to prevent further damage? Adaptation involves adjusting our lifestyles, infrastructures, and systems to minimize the damage caused by sea level rise. Mitigation, on the other hand, aims to address the underlying causes of climate change by reducing greenhouse gas emissions. This chapter delves into the ongoing debate between adaptation and mitigation, examining the arguments, strategies, and the necessity for a balanced approach.

Understanding Adaptation

Adaptation strategies focus on reducing the vulnerability of communities and ecosystems to the impacts of rising sea levels. This involves planning and implementing measures that allow societies to cope with changes and minimize harm.

1. Coastal Defenses and Infrastructure

Concept:

Coastal defenses such as sea walls, levees, and storm surge barriers are built to protect urban areas from flooding and erosion.

Examples:

The Thames Barrier in London, a movable flood barrier, protects the city from tidal surges.

The Maeslantkering in the Netherlands, a part of the Delta Works project, is one of the largest moving structures on Earth, designed to protect Rotterdam from storm surges.

"Building robust coastal defenses is crucial for protecting our cities and infrastructure from the immediate impacts of rising sea levels. However, these measures must be complemented by long-term planning and sustainable development." - Dr. John Pethick, Coastal Geomorphologist.

2. Relocation and Managed Retreat

Concept:

In some cases, relocating communities and infrastructure away from vulnerable coastal areas is necessary. Managed retreat involves strategically moving buildings and infrastructure to safer locations.

Examples:

The village of Newtok, Alaska, is being relocated due to severe coastal erosion and flooding caused by melting permafrost and rising sea levels.

In Australia, the government has implemented a managed retreat program for certain coastal areas that are highly susceptible to erosion and flooding.

"Relocating communities is a difficult but sometimes necessary decision. It requires careful planning, support for affected populations, and a commitment to creating resilient new settlements." - Dr. Elizabeth Marino, Anthropologist and Climate Migration Expert.

3. Ecosystem-Based Adaptation

Concept:

Protecting and restoring natural ecosystems, such as wetlands, mangroves, and coral reefs, can enhance coastal resilience. These ecosystems act as natural barriers, reducing the impact of storm surges and erosion.

Examples:

The restoration of mangroves in the Philippines has helped protect coastal communities from typhoons and rising sea levels.

Coral reef restoration projects in the Caribbean aim to enhance the natural defense provided by reefs against storm surges and coastal erosion.

"Ecosystem-based adaptation harnesses the power of nature to protect our coastlines. Healthy ecosystems provide vital services that enhance resilience to climate impacts." - Dr. Emily Corwin, Marine Biologist.

Understanding Mitigation

Mitigation strategies aim to address the root causes of climate change by reducing greenhouse gas emissions and enhancing carbon sinks. This approach seeks to limit the extent of climate change and its impacts, including rising sea levels.

1. Reducing Greenhouse Gas Emissions

Concept:

Cutting down emissions from fossil fuels, industrial processes, and deforestation is crucial for slowing down climate change.

Examples:

The European Union's commitment to reduce greenhouse gas emissions by at least 55% by 2030, compared to 1990 levels.

The United States rejoining the Paris Agreement and pledging to achieve net-zero emissions by 2050.

"Reducing greenhouse gas emissions is fundamental to mitigating climate change. It requires global cooperation, ambitious policies, and a transition to renewable energy sources." - Dr. James Hansen, Climate Scientist.

2. Renewable Energy and Energy Efficiency

Concept:

Transitioning to renewable energy sources such as solar, wind, and hydroelectric power, and improving energy efficiency, are key mitigation strategies.

Examples:

Germany's Energiewende (Energy Transition) initiative focuses on phasing out nuclear power and increasing the share of renewable energy in the country's energy mix.

The use of energy-efficient technologies in buildings and transportation, such as LED lighting and electric vehicles, helps reduce overall energy consumption.

"Renewable energy and energy efficiency are at the heart of climate mitigation. These strategies not only reduce emissions but also create sustainable and resilient energy systems." - Dr. Amory Lovins, Co-Founder of the Rocky Mountain Institute.

3. Carbon Sequestration and Negative Emissions Technologies

Concept:

Enhancing natural carbon sinks, such as forests and soil, and developing technologies that remove carbon dioxide from the atmosphere are important for mitigation.

Examples:

Reforestation and afforestation projects, like the Great Green Wall initiative in Africa, aim to sequester carbon and combat desertification.

Carbon capture and storage (CCS) technologies are being developed to capture CO_2 emissions from industrial processes and store them underground.

"Carbon sequestration and negative emissions technologies are essential for achieving net-zero emissions. They complement emission reduction efforts and help balance the carbon budget." - Dr. Klaus Lackner, Director of the Center for Negative Carbon Emissions.

The Debate: Adaptation vs. Mitigation

The debate between adaptation and mitigation often centers on the allocation of resources, timing, and priorities. Experts weigh in on the best approaches and the need for a balanced strategy.

4. Immediate Needs vs. Long-Term Goals

Adaptation Focus:

Proponents argue that adaptation is essential for addressing the immediate impacts of rising sea levels and protecting vulnerable communities.

Mitigation Focus:

Advocates emphasize that without mitigation, the root causes of climate change will continue to exacerbate the problem, making adaptation efforts increasingly difficult and costly.

"Adaptation is crucial for immediate protection, but without mitigation, we are merely treating the symptoms rather than the cause. Both approaches must go hand in hand." - Dr. Saleemul Huq, Director of the International Centre for Climate Change and Development.

5. Resource Allocation and Funding

Adaptation Focus:

Some argue that limited resources should be directed towards adaptation to provide immediate relief and safety for affected populations.

Mitigation Focus:

Others contend that investing in mitigation is more cost-effective in the long run, as it addresses the underlying issue and reduces future adaptation costs.

"Allocating resources wisely is key. We must ensure that both adaptation and mitigation receive adequate funding to create a comprehensive response to climate change." - Dr. Rachel Cleetus, Policy Director at the Union of Concerned Scientists.

6. Integrated Approaches and Co-Benefits

Integrated Focus:

Many experts advocate for an integrated approach that combines both adaptation and mitigation, recognizing that these strategies are complementary rather than mutually exclusive.

Co-Benefits:

Integrated strategies can provide co-benefits, such as improving air quality, creating jobs, and enhancing biodiversity, while simultaneously addressing climate change.

"An integrated approach that balances adaptation and mitigation is essential. By combining efforts, we can maximize benefits and create resilient, sustainable communities." - Dr. Kristie Ebi, Professor of Global Health and Environmental and Occupational Health Sciences.

Case Studies: Balancing Adaptation and Mitigation

7. The Netherlands: A Model of Integration

Example:

The Netherlands has long been a leader in adapting to sea level rise through its advanced flood management systems, such as the Delta Works. Simultaneously, the country is committed to reducing emissions and increasing renewable energy use.

Outcome:

This integrated approach has enabled the Netherlands to protect its population from rising sea levels while contributing to global climate mitigation efforts.

"The Netherlands demonstrates that adaptation and mitigation can go hand in hand. By investing in both areas, we can safeguard our future and contribute to global climate goals." - Henk Ovink, Special Envoy for International Water Affairs for the Kingdom of the Netherlands.

8. Bangladesh: Adaptation in Action

Example:

Bangladesh is highly vulnerable to rising sea levels and has focused heavily on adaptation strategies, such as building cyclone shelters, improving early warning systems, and implementing community-based adaptation projects.

Outcome:

These efforts have significantly reduced the loss of life and property from climate-related disasters, while the country also works on improving energy efficiency and expanding renewable energy.

"In Bangladesh, adaptation is a necessity for survival. Our efforts to protect communities must be complemented by global mitigation actions to reduce future risks." - Dr. Saleemul Huq, Director of the International Centre for Climate Change and Development.

9. California: Mitigation and Adaptation Synergies

Example:

California has implemented strong climate policies aimed at reducing greenhouse gas emissions, such as cap-and-trade programs and renewable energy mandates. At the same time, the state invests in coastal resilience projects and wildfire prevention.

Outcome:

By addressing both adaptation and mitigation, California aims to create a resilient and low-carbon future.

"California's approach shows that we can lead on both fronts. Strong climate policies and resilient infrastructure are essential for a sustainable future." - Mary Nichols, Former Chair of the California Air Resources Board.

Conclusion: Towards a Balanced Strategy

The chapter concludes by emphasizing the importance of a balanced strategy that incorporates both adaptation and mitigation. Neither approach alone is sufficient to address the complex and multifaceted challenges posed by rising sea levels and climate change. By integrating these strategies, we can protect vulnerable communities, reduce future risks, and work towards a sustainable and resilient future.

In "Adaptation vs. Mitigation," the narrative explores the critical debate between addressing the immediate impacts of rising sea levels and tackling the root causes of climate change. By highlighting expert opinions, case studies, and integrated approaches, the chapter underscores the necessity of a comprehensive strategy that balances adaptation and mitigation efforts. This discussion sets the stage for exploring innovative solutions, international cooperation, and the role of communities in subsequent chapters, further highlighting the multifaceted nature of the challenge.

Chapter 14
The Insurance Dilemma

Introduction: Rising Risks, Rising Premiums

The insurance industry is a critical player in the response to rising sea levels, as it provides financial protection against natural disasters and climate-related damages. However, as sea levels rise and the frequency of extreme weather events increases, insurers are facing unprecedented challenges. This chapter delves into the insurance industry's response to the growing risks posed by rising sea levels, exploring changes in policy, premiums, and coverage. It also examines the challenges faced by homeowners and businesses, and the evolving landscape of risk management in a climate-impacted world.

The Growing Risk Landscape

Rising sea levels increase the risk of flooding, storm surges, and coastal erosion, leading to significant financial losses for individuals, businesses, and insurers. Understanding this evolving risk landscape is essential for the insurance industry to adapt and respond effectively.

1. Increased Frequency and Severity of Flooding

Concept:

As sea levels rise, the frequency and severity of coastal flooding events are expected to increase. This not only causes direct damage to properties but also disrupts communities and economies.

Examples:

The record-breaking floods in Venice in 2019 caused extensive damage to homes, businesses, and historic sites, resulting in substantial insurance claims.

The aftermath of Hurricane Sandy in 2012 highlighted the vulnerability of coastal cities like New York to storm surges, leading to billions of dollars in insured losses.

"The increasing frequency and severity of flooding due to rising sea levels present significant challenges for the insurance industry. We must adapt our models and policies to reflect this new reality." - Dr. Robert Muir-Wood, Chief Research Officer at RMS.

2. Coastal Erosion and Property Damage

Concept:

Coastal erosion, exacerbated by rising sea levels, leads to the gradual loss of land and damage to properties. This creates long-term risks for coastal communities and infrastructure.

Examples:

In Norfolk, England, homes have been lost to the sea due to accelerated coastal erosion, leading to significant financial losses for property owners and insurers.

The Outer Banks in North Carolina face ongoing erosion, threatening residential properties and tourism infrastructure, with insurance companies adjusting their risk assessments accordingly.

"Coastal erosion poses a long-term threat to property and infrastructure. Insurers need to account for these gradual changes and the cumulative impacts on communities." - Dr. Orrin Pilkey, Professor Emeritus of Earth and Ocean Sciences at Duke University.

Insurance Industry Responses

The insurance industry is adapting to the growing risks posed by rising sea levels through changes in policy, premiums, and coverage. These adaptations aim to manage risk, ensure solvency, and continue providing coverage to affected communities.

3. Adjusting Premiums and Coverage

Concept:

Insurers are adjusting premiums and coverage terms to reflect the increased risk of flooding and coastal erosion. This often results in higher costs for policyholders in vulnerable areas.

Examples:

In Florida, homeowners in coastal areas have seen significant increases in insurance premiums due to the heightened risk of hurricanes and flooding.

The National Flood Insurance Program (NFIP) in the United States has reformed its pricing structure to more accurately reflect the risk of flooding, leading to higher premiums for some policyholders.

"Adjusting premiums and coverage terms is a necessary step to ensure the viability of the insurance industry in the face of rising sea levels. However, this also raises concerns about affordability and accessibility for homeowners." - David Maurstad, Chief Executive of the NFIP.

4. Exclusions and Limitations

Concept:

Some insurers are introducing exclusions and limitations on coverage for flood and coastal erosion damage, reducing their exposure to high-risk claims.

Examples:

In Australia, certain insurance policies now exclude coverage for damage caused by gradual coastal erosion, leaving homeowners to bear the financial burden.

In the UK, insurers have introduced limitations on flood coverage for properties in high-risk areas, prompting homeowners to seek alternative risk management solutions.

"Exclusions and limitations on coverage are measures that insurers use to manage their risk exposure. However, this can leave policyholders without adequate protection against the impacts of rising sea levels." - Amanda Blanc, Group CEO of Aviva.

5. Innovative Risk Management Solutions

Concept:

The insurance industry is developing innovative risk management solutions to address the challenges posed by rising sea levels. These include parametric insurance, risk pooling, and public-private partnerships.

Examples:

Parametric insurance, which pays out based on predefined triggers such as specific flood levels, is being explored as a way to provide quicker and more predictable payouts to policyholders.

The "Flood Re" program in the UK is a public-private partnership that helps insurers provide affordable flood insurance to high-risk properties by pooling risk and subsidizing premiums.

"Innovative risk management solutions, such as parametric insurance and risk pooling, offer new ways to address the financial challenges of rising sea levels. These approaches can enhance resilience and provide more reliable coverage for policyholders." - Rowan Douglas, Head of Climate and Resilience Hub at Willis Towers Watson.

Challenges for Homeowners and Businesses

Homeowners and businesses in coastal areas face significant challenges as they navigate the evolving insurance landscape. These challenges include increased costs, reduced coverage options, and the need for proactive risk management.

6. Affordability and Access to Insurance

Concept:

As premiums rise and coverage becomes more limited, affordability and access to insurance become major concerns for homeowners and businesses in high-risk areas.

Examples:

In Louisiana, many homeowners are struggling to afford rising flood insurance premiums, leading to a decrease in the number of insured properties.

Small businesses in coastal areas are finding it difficult to secure affordable insurance coverage, affecting their ability to operate and recover from climate-related events.

"The rising cost of insurance and reduced coverage options pose significant challenges for homeowners and businesses. We need to find ways to ensure that insurance remains accessible and affordable for those at risk." - Roy Wright, President & CEO of the Insurance Institute for Business & Home Safety.

7. Proactive Risk Management and Adaptation

Concept:

Homeowners and businesses are increasingly encouraged to adopt proactive risk management and adaptation measures to reduce their vulnerability to rising sea levels and enhance their insurability.

Examples:

Elevating homes and implementing flood-resistant construction techniques can help reduce flood risk and lower insurance premiums for homeowners in vulnerable areas.

Businesses are investing in resilient infrastructure and continuity planning to minimize disruptions and ensure they can recover quickly from climate-related events.

"Proactive risk management and adaptation measures are crucial for reducing vulnerability and maintaining insurability in the face of rising sea levels. By taking these steps, homeowners and businesses can enhance their resilience and protect their investments." - Dr. Carolyn Kousky, Executive Director of the Wharton Risk Management and Decision Processes Center.

8. Government Support and Policy Interventions

Concept:

Government support and policy interventions are essential for addressing the insurance challenges posed by rising sea levels. This includes subsidies, regulatory reforms, and investment in resilience infrastructure.

Examples:

The US government provides subsidies through the NFIP to help homeowners afford flood insurance and invests in flood mitigation projects to reduce risk.

In the Netherlands, the government works closely with insurers and communities to implement comprehensive flood protection measures, ensuring that insurance remains viable and accessible.

"Government support and policy interventions play a crucial role in addressing the insurance challenges of rising sea levels. By working together, we can create a more resilient and sustainable future for vulnerable communities." - Dr. Michel Kerres, Senior Advisor on Water Management at the Dutch Ministry of Infrastructure and Water Management.

Conclusion: Navigating the Insurance Dilemma

The chapter concludes by emphasizing the importance of collaboration between the insurance industry, policymakers, homeowners, and businesses in navigating the challenges posed by rising sea levels. By adjusting premiums and coverage, developing innovative risk management solutions, and adopting proactive adaptation measures, we can create a more resilient and sustainable insurance landscape. Ensuring that insurance remains accessible and affordable for those at risk is essential for protecting communities and supporting recovery efforts in the face of a changing climate.

In "The Insurance Dilemma," the narrative explores the complex interplay between rising sea levels and the insurance industry. By highlighting the industry's responses, the challenges faced by policyholders, and the need for proactive risk management, the chapter underscores the critical role of insurance in building resilience and supporting adaptation efforts.

This discussion sets the stage for exploring the broader economic and social implications of rising sea levels in subsequent chapters, further highlighting the multifaceted nature of the challenge.

Chapter 15
Legal and Ethical Considerations

Introduction: Navigating Complex Waters

As rising sea levels threaten coastal cities worldwide, legal and ethical considerations become increasingly crucial. This chapter explores the intricate legal issues surrounding property rights, compensation, and relocation, as well as the ethical questions about responsibility and the costs of addressing rising sea levels. Through a detailed analysis of case studies, expert opinions, and theoretical frameworks, this chapter sheds light on the challenges and dilemmas faced by individuals, communities, governments, and the international community in responding to this pressing issue.

Legal Issues in Property Rights and Compensation

Rising sea levels create significant legal challenges related to property rights and compensation. As land is lost to the sea and properties become uninhabitable, determining ownership, liability, and compensation becomes a complex and contentious issue.

1. Property Rights and Land Loss

Concept:

Rising sea levels lead to the gradual erosion of coastlines and the loss of land, raising questions about property rights and ownership.

Examples:

In the United States, coastal erosion has led to disputes over property boundaries and ownership as land disappears or becomes submerged.

Pacific island nations, such as Kiribati and Tuvalu, face existential threats as rising sea levels encroach on their territory, raising international legal questions about sovereignty and territorial rights.

"The loss of land due to rising sea levels presents unprecedented challenges for property rights and ownership. Legal frameworks must adapt to address these emerging issues." - Dr. Robin Kundis Craig, Professor of Law at the University of Utah.

2. Compensation for Damages

Concept:

As properties are damaged or rendered uninhabitable by rising sea levels, the issue of compensation becomes critical. Determining who is responsible for compensating affected individuals and communities is a complex legal challenge.

Examples:

After Hurricane Katrina, homeowners in New Orleans faced significant legal battles over insurance claims and compensation for flood damage, highlighting the difficulties in obtaining fair compensation.

In Bangladesh, the government has initiated programs to compensate families displaced by river erosion, but these efforts are often hampered by limited resources and legal complexities.

"Compensating individuals and communities for the loss and damage caused by rising sea levels is a significant legal and ethical challenge. Ensuring fair and just compensation is essential for supporting those affected." - Dr. Michael Gerrard, Director of the Sabin Center for Climate Change Law at Columbia Law School.

3. Relocation and Managed Retreat

Concept:

As some areas become uninhabitable due to rising sea levels, relocation and managed retreat become necessary strategies. Legal frameworks must address the complexities of relocating communities and compensating displaced individuals.

Examples:

In Alaska, the village of Newtok is being relocated due to severe erosion and flooding. Legal and logistical challenges have complicated the relocation process, highlighting the need for comprehensive legal frameworks.

In Indonesia, the government has announced plans to relocate the capital city from Jakarta to Borneo due to subsidence and rising sea levels, raising complex legal issues related to land acquisition and compensation.

"Relocating communities is a complex legal and logistical challenge. Comprehensive legal frameworks are needed to ensure that relocation is conducted fairly and that affected individuals are adequately compensated." - Dr. Daniel Farber, Professor of Law at the University of California, Berkeley.

Ethical Questions of Responsibility and Cost

The ethical questions surrounding who bears the responsibility and costs of addressing rising sea levels are deeply intertwined with issues of justice, equity, and human rights. These questions are critical in shaping policies and responses to the impacts of climate change.

4. Responsibility for Climate Change

Concept:

Determining who is responsible for the greenhouse gas emissions that drive climate change and rising sea levels is a contentious ethical issue. This involves questions of historical responsibility, industrial development, and global equity.

Examples:

Industrialized nations, which have historically contributed the most to greenhouse gas emissions, face ethical questions about their responsibility to support adaptation and mitigation efforts in developing countries.

The principle of "common but differentiated responsibilities" under international climate agreements reflects the ethical recognition that different countries have different levels of responsibility and capacity to address climate change.

"The ethical question of responsibility for climate change is complex and deeply rooted in issues of historical emissions and global equity. We must acknowledge these ethical dimensions in our responses to rising sea levels." - Dr. Henry Shue, Professor of Politics and International Relations at the University of Oxford.

5. Equity and Justice in Adaptation and Mitigation

Concept:

Ensuring that adaptation and mitigation efforts are equitable and just is a fundamental ethical consideration. This involves addressing the needs and rights of vulnerable and marginalized communities.

Examples:

In the United States, low-income and minority communities are often disproportionately affected by climate impacts, raising ethical questions about how adaptation efforts can address these inequities.

Indigenous communities in the Arctic, who contribute the least to global emissions, are among the most affected by rising sea levels, highlighting the need for ethical and just responses that respect their rights and ways of life.

"Equity and justice must be at the heart of our adaptation and mitigation efforts. We must ensure that the most vulnerable and marginalized communities are not left behind." - Dr. Sheila Jasanoff, Professor of Science and Technology Studies at Harvard University.

6. The Cost of Adaptation and Mitigation

Concept:

The financial costs of adaptation and mitigation are significant, and determining who should bear these costs is a major ethical issue. This involves questions of funding, resource allocation, and global cooperation.

Examples:

The Green Climate Fund was established to support developing countries in their efforts to adapt to and mitigate climate change, reflecting an ethical commitment to shared global responsibility.

In the Maldives, the government is investing heavily in adaptation measures such as building sea walls and elevating infrastructure, raising questions about the financial burden on small island nations with limited resources.

"The costs of adaptation and mitigation are substantial, and determining who should bear these costs is an ethical challenge. We must ensure that financial resources are allocated fairly and equitably." - Dr. Saleemul Huq, Director of the International Centre for Climate Change and Development.

Case Studies: Legal and Ethical Dilemmas in Action

7. The Marshall Islands: Legal Battles for Survival

Example: The Marshall Islands, a low-lying Pacific nation, faces existential threats from rising sea levels. The government has engaged in international legal battles, seeking accountability and compensation from major greenhouse gas emitters.

Outcome:

These legal efforts highlight the challenges of addressing climate justice and the ethical responsibility of industrialized nations to support vulnerable countries.

"The Marshall Islands' legal battles underscore the urgency of addressing climate justice and the ethical responsibility of industrialized nations to support those most affected by rising sea levels." - Dr. Tony de Brum, Former Foreign Minister of the Marshall Islands.

8. Miami, Florida: Balancing Development and Risk

Example:

Miami faces significant risks from rising sea levels and frequent flooding. The city's efforts to balance ongoing development with adaptation measures raise complex legal and ethical questions about property rights, compensation, and the responsibilities of developers and policymakers.

Outcome:

Miami's experience highlights the need for comprehensive legal frameworks and ethical considerations in urban planning and development in the face of rising sea levels.

"Miami's efforts to balance development and risk illustrate the complex legal and ethical challenges of urban planning in the context of rising sea levels. We must ensure that development is sustainable and equitable." - Dr. Jayantha

Obeysekera, Director of the Sea Level Solutions Center at Florida International University.

9. The Sundarbans, India and Bangladesh: Shared Responsibility

Example:

The Sundarbans, a vast mangrove forest spanning India and Bangladesh, is threatened by rising sea levels and increased salinity. The shared responsibility of the two countries to protect this unique ecosystem raises legal and ethical questions about cross-border cooperation and resource allocation.

Outcome:

The Sundarbans case underscores the Importance of international cooperation and ethical responsibility in addressing shared environmental challenges.

"The Sundarbans highlight the need for cross-border cooperation and shared ethical responsibility in addressing the impacts of rising sea levels. Protecting this unique ecosystem requires a collective effort." - Dr. Sugata Hazra, Professor of Oceanography at Jadavpur University.

Conclusion: Towards Ethical and Legal Solutions

The chapter concludes by emphasizing the importance of addressing the legal and ethical considerations in responding to rising sea levels. Comprehensive legal frameworks and ethical principles must guide our efforts to protect vulnerable communities, ensure fair compensation, and allocate resources equitably. By recognizing the complexities and dilemmas involved, we can work towards solutions that are just, sustainable, and effective.

In "Legal and Ethical Considerations," the narrative explores the intricate legal and ethical challenges posed by rising sea levels. By highlighting property rights, compensation, relocation, and the broader questions of responsibility and cost, the chapter underscores the need for a holistic approach that integrates legal and ethical dimensions. This discussion sets the stage for exploring the roles of technology, innovation, and international cooperation in subsequent chapters, further highlighting the multifaceted nature of the challenge.

Chapter 16
Cultural Heritage at Risk

Introduction: The Looming Threat to Cultural Heritage

Rising sea levels pose a significant threat not only to the physical landscapes of coastal cities but also to their cultural heritage. Historic landmarks, heritage sites, and cultural treasures that have stood for centuries are now at risk of being lost to the encroaching waters. This chapter delves into the impact of rising sea levels on cultural heritage, exploring efforts to preserve these treasures and emphasizing the importance of cultural continuity for future generations.

The Cultural Heritage Under Threat

Coastal cities around the world are home to a wealth of cultural landmarks and heritage sites that are integral to their identity and history. As sea levels rise, these irreplaceable assets face a variety of risks, including flooding, erosion, and structural damage.

1. Historic Landmarks and Buildings

Concept:

Historic buildings and landmarks, often located in coastal areas, are particularly vulnerable to rising sea levels and associated flooding.

Examples:

Venice, Italy: Known for its historic architecture and canals, Venice faces frequent flooding due to rising sea levels. Iconic landmarks like St. Mark's Basilica are at risk of water damage.

Alexandria, Egypt: The ancient city, home to numerous archaeological sites, is threatened by coastal erosion and rising sea levels, endangering its rich cultural history.

"Historic landmarks are more than just buildings; they are tangible connections to our past. Protecting them from rising sea levels is crucial for preserving our cultural heritage." - Dr. Antonia Layard, Professor of Law and Geography at the University of Bristol.

2. Archaeological Sites

Concept:

Coastal archaeological sites provide invaluable insights into human history and prehistoric civilizations, but rising sea levels threaten to submerge and destroy these sites.

Examples:

The ruins of the ancient city of Heracleion, submerged off the coast of Egypt, have been further threatened by rising sea levels, complicating archaeological efforts.

The Jomon Sannai-Maruyama site in Japan, an important prehistoric settlement, faces risks from rising sea levels and increased flooding, which could erode the site.

"Archaeological sites offer a glimpse into our collective history. Rising sea levels threaten to erase these invaluable sources of knowledge forever." - Dr. Sarah Parcak, Professor of Anthropology at the University of Alabama.

3. Cultural Landscapes

Concept:

Cultural landscapes, including historic gardens, parks, and natural sites with cultural significance, are also at risk. These landscapes contribute to the identity and heritage of communities.

Examples:

The Everglades in Florida, a UNESCO World Heritage Site, is both a natural and cultural landscape, home to indigenous histories and diverse ecosystems, now threatened by rising sea levels and saltwater intrusion.

The coastal rice terraces of the Philippine Cordilleras, a cultural landscape and UNESCO World Heritage Site, are vulnerable to climate change-induced sea level rise, affecting traditional farming practices and community life.

"Cultural landscapes are living testimonies to the relationship between people and their environment. Protecting these landscapes is essential for maintaining cultural continuity." - Dr. Richard Longstreth, Professor of American Studies at George Washington University.

Efforts to Preserve Cultural Heritage

Efforts to preserve cultural heritage in the face of rising sea levels involve a combination of innovative technologies, community initiatives, and policy measures. These efforts aim to protect and adapt heritage sites while maintaining their cultural significance.

4. Technological Innovations in Preservation

Concept:

Advanced technologies are being utilized to protect and document cultural heritage sites threatened by rising sea levels.

Examples:

Digital documentation and 3D modeling: Projects like CyArk are creating digital records of endangered heritage sites, enabling virtual preservation and future restoration efforts.

Protective engineering: The MOSE project in Venice involves a series of movable barriers designed to protect the city from flooding and rising sea levels.

"Innovative technologies offer new ways to preserve and protect our cultural heritage from the impacts of rising sea levels. These tools are essential for future generations to access and appreciate their history." - Dr. Elizabeth Lee, CEO of CyArk.

5. Community-Based Initiatives

Concept:

Local communities play a vital role in preserving cultural heritage. Community-based initiatives focus on raising awareness, promoting local stewardship, and integrating traditional knowledge in preservation efforts.

Examples:

In Fiji, community-led projects are working to protect coastal heritage sites through traditional practices and sustainable management, emphasizing the role of indigenous knowledge.

The historic city of Annapolis, Maryland, has involved local communities in efforts to adapt and protect heritage sites from flooding through public workshops and engagement.

"Community-based initiatives are crucial for the preservation of cultural heritage. Empowering local communities ensures that preservation efforts are sustainable and culturally sensitive." - Dr. Jane Lennon, Heritage Consultant and UNESCO Advisor.

6. Policy Measures and International Cooperation

Concept:

Governments and international organizations are implementing policy measures and fostering cooperation to protect cultural heritage from rising sea levels.

Examples:

UNESCO's World Heritage Centre is actively involved in assessing the risks to heritage sites and developing strategies for their protection, including guidance on climate adaptation for heritage management.

National policies: Countries like the Netherlands have integrated cultural heritage protection into their national climate adaptation plans, ensuring that heritage sites are considered in broader resilience strategies.

"Effective policy measures and international cooperation are essential for safeguarding our cultural heritage. Collaborative efforts can enhance resilience and ensure that heritage sites are preserved for future generations." - Dr. Mechtild Rössler, Director of the UNESCO World Heritage Centre.

The Importance of Cultural Continuity

Preserving cultural heritage in the face of rising sea levels is not just about protecting physical structures and sites; it is also about maintaining cultural continuity and identity. Cultural heritage connects communities to their past and provides a sense of identity and belonging.

7. Cultural Identity and Community Resilience

Concept:

Cultural heritage is integral to the identity of communities. Preserving heritage sites contributes to the resilience and continuity of cultural traditions and practices.

Examples:

In the Solomon Islands, the preservation of traditional fishing practices and cultural sites is seen as essential for community resilience and cultural identity in the face of rising sea levels.

Coastal cities like Charleston, South Carolina, are working to protect historic districts and cultural landmarks, recognizing their importance to the city's identity and tourism economy.

"Cultural heritage is a cornerstone of community identity and resilience. Protecting it is essential for ensuring that cultural traditions and practices continue to thrive." - Dr. Laurajane Smith, Professor of Heritage Studies at the Australian National University.

8. Intergenerational Knowledge and Legacy

Concept:

Preserving cultural heritage ensures that future generations can access, learn from, and appreciate their history and cultural legacy.

Examples:

The preservation of indigenous heritage sites in Alaska, threatened by coastal erosion, ensures that future generations can continue to learn from and connect with their cultural legacy.

Educational programs and heritage conservation projects in places like the UK's Norfolk Coast aim to pass on knowledge and appreciation of cultural heritage to younger generations.

"Ensuring that future generations have access to their cultural heritage is a crucial aspect of our responsibility. It allows them to understand their history, identity, and place in the world." - Dr. David Lowenthal, Historian and Geographer.

Conclusion: Safeguarding Our Cultural Legacy

The chapter concludes by emphasizing the critical importance of safeguarding cultural heritage in the face of rising sea levels. Through technological innovations, community initiatives, policy measures, and international cooperation, we can protect these irreplaceable treasures and ensure the continuity of cultural identity and legacy for future generations.

In "Cultural Heritage at Risk," the narrative explores the profound impact of rising sea levels on cultural landmarks and heritage sites. By highlighting the efforts to preserve these treasures and emphasizing the importance of cultural continuity, the chapter underscores the need for a comprehensive approach to heritage conservation in the context of climate change. This discussion sets the stage for exploring the roles of communities, governments, and international organizations in the broader effort to build resilience and adapt to rising sea levels in subsequent chapters.

Chapter 17
Youth Voices and Future Leaders

Introduction: The Rise of Youth Activism

As the world grapples with the impacts of rising sea levels, young activists and future leaders are emerging as powerful voices in the fight against climate change. This chapter delves into the perspectives and initiatives of young people committed to combating rising sea levels. Highlighting their innovative approaches and relentless activism, it underscores the critical role of the next generation in driving meaningful change and shaping a sustainable future.

The Power of Youth Activism

Across the globe, young people are stepping up to address the challenges posed by rising sea levels. Their activism is characterized by a sense of urgency, creativity, and a deep commitment to environmental justice.

1. Global Youth Movements

Concept:

Youth movements have gained international prominence, mobilizing millions and influencing climate policies.

Examples:

Greta Thunberg and the Fridays for Future movement have galvanized youth around the world to demand urgent action on climate change. Their protests emphasize the need for immediate measures to address rising sea levels and other climate impacts.

The Sunrise Movement in the United States advocates for a Green New Deal, aiming to combat climate change and create jobs. Their focus includes addressing coastal resilience and protecting communities from rising sea levels.

"We deserve a safe future. And we demand a safe future. Is that really too much to ask?" - Greta Thunberg, Climate Activist.

2. Local Activism and Community Engagement

Concept:

Young activists are driving change within their communities, focusing on local issues and solutions.

Examples:

In the Philippines, young activists are organizing coastal cleanups and mangrove planting projects to protect shorelines from erosion and rising sea levels. Their efforts combine environmental conservation with community education.

In the UK, young climate leaders like Ella and Caitlin McEwan successfully campaigned to ban single-use plastics in schools, reducing pollution that exacerbates coastal flooding and sea level rise.

"By taking action locally, we can make a difference globally. Every small step contributes to a larger movement for change." - Licypriya Kangujam, Youth Climate Activist.

Innovating for a Sustainable Future

The next generation is not only advocating for change but also driving innovation. Young entrepreneurs and scientists are developing creative solutions to mitigate and adapt to rising sea levels.

3. Technological Innovation

Concept:

Young innovators are leveraging technology to develop solutions for climate resilience and adaptation.

Examples:

Boyan Slat, a Dutch inventor, founded The Ocean Cleanup project at age 18. His initiative aims to remove plastic from the oceans, reducing pollution and protecting marine ecosystems that buffer against rising sea levels.

Gitanjali Rao, a young scientist from the United States, developed a device to detect lead in water, which can be adapted for monitoring water quality in coastal areas affected by rising sea levels.

"Innovation is key to solving the climate crisis. By harnessing technology, we can create sustainable solutions that protect our planet." - Boyan Slat, Founder of The Ocean Cleanup.

4. Sustainable Entrepreneurship

Concept:

Young entrepreneurs are creating businesses that address climate challenges while promoting sustainability.

Examples:

Fionn Ferreira, an Irish entrepreneur, developed a method to remove microplastics from water, helping to protect marine life and coastal ecosystems from pollution exacerbated by rising sea levels.

Maanasa Mendu, a young inventor, created a cost-effective device that harnesses wind and solar energy, providing renewable energy solutions for coastal communities vulnerable to rising sea levels.

"Entrepreneurship allows us to turn ideas into impactful solutions. By focusing on sustainability, we can build a future that is resilient and thriving." - Fionn Ferreira, Environmental Scientist and Inventor.

Education and Empowerment

Education plays a crucial role in empowering young people to address the challenges of rising sea levels. By equipping them with knowledge and skills, we can foster a generation of informed and proactive leaders.

5. Climate Education Programs

Concept: Educational initiatives are essential for raising awareness and building the capacity of young people to tackle climate issues.

Examples:

The Climate Reality Project, founded by former Vice President Al Gore, offers training programs for young people, empowering them with the knowledge and tools to advocate for climate action, including addressing sea level rise.

The International Coastal Cleanup, organized by the Ocean Conservancy, engages young volunteers in hands-on conservation efforts, educating them about the impacts of marine debris and rising sea levels.

"Education is the foundation for action. By empowering young people with knowledge, we can inspire them to become leaders in the fight against climate change." - Al Gore, Founder of The Climate Reality Project.

6. School and University Initiatives

Concept:

Schools and universities are integrating climate education into their curricula and encouraging student-led sustainability projects.

Examples:

The University of the South Pacific offers courses on climate change and adaptation, focusing on the unique challenges faced by Pacific island nations. Students are actively involved in research and community outreach projects related to rising sea levels.

High schools in coastal areas are implementing sustainability programs, such as the Green Schools Alliance, which encourages students to develop projects that address local climate impacts and promote environmental stewardship.

"Educational institutions have a critical role in preparing the next generation to tackle climate challenges. By fostering a culture of sustainability, we can equip students with the skills they need to make a difference." - Dr. Jale Samuwai, Lecturer at the University of the South Pacific.

The Role of Youth in Policy and Advocacy

Young people are increasingly involved in policy and advocacy efforts, pushing for legislative changes and holding leaders accountable for climate action.

7. Youth in Policy-Making

Concept:

Young activists are participating in policy-making processes, bringing their perspectives and demands to the forefront of climate discussions.

Examples:

At the United Nations Climate Change Conference (COP), youth delegates from around the world advocate for stronger climate commitments, including measures to address rising sea levels. Their participation ensures that youth voices are heard in international policy discussions.

In the United States, youth-led organizations like Zero Hour are lobbying for comprehensive climate policies, including adaptation strategies for coastal communities threatened by rising sea levels.

"Youth involvement in policy-making is essential for driving ambitious climate action. We must ensure that our voices are heard and our future is protected." - Jamie Margolin, Co-Founder of Zero Hour.

8. Holding Leaders Accountable

Concept:

Young activists are holding leaders accountable for their climate commitments, demanding transparency and action.

Examples:

In Australia, youth activists from the School Strike 4 Climate movement have organized mass protests and taken legal action to hold the government accountable for its climate policies, highlighting the urgency of addressing sea level rise.

In Canada, the youth-led lawsuit La Rose et al. v. Her Majesty the Queen argues that the government's inaction on climate change violates their rights to life, liberty, and security. The case emphasizes the need for robust measures to protect coastal communities from rising sea levels.

"We must hold our leaders accountable for their climate commitments. Our future depends on their actions today." - Anjali Appadurai, Climate Justice Activist.

Conclusion: The Future is in Their Hands

The chapter concludes by emphasizing the critical role of youth in combating rising sea levels and shaping a sustainable future. Through activism, innovation, education, and advocacy, young people are driving the change needed to address the climate crisis. By supporting and empowering the next generation, we can ensure that they have the tools and opportunities to build a resilient and sustainable world.

In "Youth Voices and Future Leaders," the narrative explores the dynamic and inspiring contributions of young people to the fight against rising sea levels. By highlighting their perspectives, initiatives, and unwavering commitment to climate action, the chapter underscores the importance of youth leadership in addressing one of the most pressing challenges of our time. This discussion sets the stage for exploring the broader implications of intergenerational cooperation and global solidarity in subsequent chapters, further emphasizing the need for a united and inclusive approach to combating rising sea levels.

Chapter 18
The Role of Industry

Introduction: Industry's Impact on Rising Sea Levels

Industries worldwide significantly influence climate change, contributing to the rising sea levels threatening coastal communities. This chapter explores the impact of various industries on rising sea levels and their efforts to reduce their carbon footprint. From fossil fuels to fashion, we examine how different sectors are responding to the crisis and what measures they are taking to mitigate their environmental impact.

The Fossil Fuel Industry

The fossil fuel industry is a major contributor to greenhouse gas emissions, driving climate change and consequently, rising sea levels. However, there are emerging efforts within the sector to shift towards more sustainable practices.

1. Impact of Fossil Fuels

Concept:

The extraction, production, and consumption of fossil fuels release significant amounts of carbon dioxide (CO_2) and methane (CH_4) into the atmosphere.

Examples:

Coal-fired power plants are one of the largest sources of CO_2 emissions globally, contributing to global warming and rising sea levels.

Oil and gas extraction, particularly from offshore drilling, poses risks of spills and leaks, exacerbating environmental damage.

"The burning of fossil fuels is the single largest contributor to global warming. Transitioning to renewable energy is critical for mitigating rising sea levels." - Dr. James Hansen, Climate Scientist.

2. Efforts to Reduce Carbon Footprint

Concept:

Despite its significant impact, the fossil fuel industry is beginning to adopt measures to reduce emissions and transition to cleaner energy sources.

Examples:

Companies like BP and Shell have announced net-zero emissions targets and are investing in renewable energy projects, such as wind and solar farms.

ExxonMobil and Chevron are developing carbon capture and storage (CCS) technologies to reduce CO_2 emissions from their operations.

"While progress is being made, the fossil fuel industry must accelerate its efforts to reduce emissions and invest in sustainable energy solutions." - Christiana Figueres, Former Executive Secretary of the UNFCCC.

The Manufacturing and Construction Industries

Manufacturing and construction are significant contributors to greenhouse gas emissions, largely due to their reliance on energy-intensive processes and materials.

3. Impact of Manufacturing

Concept:

The manufacturing sector, including steel, cement, and chemical production, is energy-intensive and heavily reliant on fossil fuels.

Examples:

Cement production accounts for approximately 8% of global CO_2 emissions, due to the energy required to produce clinker, the main ingredient in cement.

The steel industry is another major emitter, with CO_2 released during the smelting process and from the combustion of fossil fuels in blast furnaces.

"Reducing emissions from manufacturing requires innovative technologies and a commitment to sustainable practices across the industry." - Dr. Jeffrey Sachs, Economist and Sustainable Development Expert.

4. Sustainable Practices and Innovations

Concept:

The manufacturing sector is adopting various sustainable practices to reduce its environmental impact.

Examples:

Green cement technologies, such as using alternative materials like fly ash and slag, are being developed to reduce CO_2 emissions from cement production.

The steel industry is exploring hydrogen-based steelmaking processes, which significantly lower emissions compared to traditional methods.

"Innovation in manufacturing is key to achieving sustainability. By adopting green technologies, the industry can reduce its carbon footprint and contribute to mitigating rising sea levels." - Dr. Fatih Birol, Executive Director of the International Energy Agency.

5. Impact of Construction

Concept:

The construction industry contributes to emissions through the energy required for building materials, transportation, and construction activities.

Examples:

The construction of buildings and infrastructure consumes large amounts of energy and materials, leading to significant CO_2 emissions.

Urban sprawl and deforestation for construction projects exacerbate climate change by reducing natural carbon sinks.

"Sustainable construction practices are essential for reducing emissions and building resilience against rising sea levels." - Dr. Peter Graham, Executive Director of the Global Buildings Performance Network.

6. Green Building Initiatives

Concept:

The construction industry is increasingly adopting green building practices to minimize its environmental impact.

Examples:

The Leadership in Energy and Environmental Design (LEED) certification promotes sustainable building practices, including energy efficiency, water conservation, and the use of eco-friendly materials.

Passive House standards focus on creating highly energy-efficient buildings that require minimal heating and cooling, reducing overall emissions.

"Green building initiatives not only reduce emissions but also enhance the resilience of communities to the impacts of rising sea levels." - Rick Fedrizzi, Co-Founder of the U.S. Green Building Council.

The Fashion Industry

The fashion industry, known for its fast-paced production cycles, also contributes to climate change through its extensive use of resources and generation of waste.

7. Impact of Fast Fashion

Concept:

Fast fashion relies on quick production and low-cost materials, resulting in high levels of waste and emissions.

Examples:

The fashion industry is responsible for around 10% of global carbon emissions, primarily due to the energy-intensive production of textiles and the transportation of goods.

Textile dyeing and treatment processes release harmful chemicals into water bodies, affecting marine ecosystems and contributing to environmental degradation.

"The fast fashion model is inherently unsustainable. We need to shift towards a circular economy in the fashion industry to mitigate its impact on rising sea levels." - Stella McCartney, Fashion Designer and Sustainability Advocate.

8. Sustainable Fashion Practices

Concept:

The fashion industry is increasingly embracing sustainability through various initiatives and practices.

Examples:

Brands like Patagonia and Eileen Fisher are leading the way in sustainable fashion, using recycled materials, organic cotton, and ethical production methods.

The rise of slow fashion promotes quality over quantity, encouraging consumers to invest in durable, timeless pieces rather than disposable trends.

"Sustainable fashion is not just a trend; it's a necessity. By making conscious choices, we can reduce the industry's environmental impact and contribute to a more sustainable future." - Eileen Fisher, Fashion Designer.

The Role of the Technology Sector

The technology sector, while contributing to emissions through energy consumption, also offers solutions for reducing carbon footprints and mitigating climate impacts.

9. Impact of Technology

Concept:

The tech industry, particularly data centers and manufacturing of electronic devices, consumes significant energy and resources.

Examples:

Data centers, essential for cloud computing and digital services, require vast amounts of electricity and cooling, contributing to CO_2 emissions.

The production of electronic devices, such as smartphones and computers, involves mining for rare earth metals and energy-intensive manufacturing processes.

"The tech sector must address its environmental impact by adopting energy-efficient practices and supporting renewable energy." - Sundar Pichai, CEO of Alphabet Inc.

10. Technological Solutions for Sustainability

Concept:

Technology companies are leveraging innovation to develop sustainable solutions and reduce their carbon footprint.

Examples:

Google and Apple have committed to using 100% renewable energy for their data centers and operations, significantly reducing their carbon emissions.

Companies like Tesla are advancing electric vehicle technology, providing alternatives to fossil fuel-powered transportation and reducing overall emissions.

"Technology has the potential to drive significant progress in sustainability. By embracing clean energy and innovative solutions, we can mitigate the impacts of rising sea levels." - Elon Musk, CEO of Tesla.

The Agricultural Sector

Agriculture is a major contributor to greenhouse gas emissions, particularly methane from livestock and nitrous oxide from fertilizers. However, sustainable farming practices can help mitigate these impacts.

11. Impact of Agriculture

Concept:

Agricultural practices contribute to climate change through deforestation, methane emissions from livestock, and the use of synthetic fertilizers.

Examples:

Livestock farming, particularly cattle, produces large amounts of methane, a potent greenhouse gas that contributes to global warming and rising sea levels.

Deforestation for agricultural expansion reduces the planet's capacity to absorb CO_2, exacerbating climate change.

"Agriculture is a significant driver of climate change, but it also has the potential to be part of the solution through sustainable practices." - Dr. Jonathan Foley, Executive Director of Project Drawdown.

12. Sustainable Agriculture Practices

Concept:

Sustainable farming practices aim to reduce emissions, enhance soil health, and promote biodiversity.

Examples:

Regenerative agriculture focuses on soil health, crop diversity, and minimal tillage, improving carbon sequestration and reducing emissions.

Agroforestry integrates trees and shrubs into farming systems, enhancing carbon storage, protecting against soil erosion, and promoting biodiversity.

"Sustainable agriculture is essential for mitigating climate change and ensuring food security. By adopting regenerative practices, we can build resilience and reduce emissions." - Dr. Vandana Shiva, Environmental Activist and Author.

Conclusion: Industry's Role in Combating Rising Sea Levels

The chapter concludes by emphasizing the critical role of various industries in addressing the challenges of rising sea levels. While the impact of these sectors on climate change is significant, their efforts to adopt sustainable practices and reduce emissions are crucial for mitigating the crisis. By embracing innovation, committing to sustainability, and collaborating across sectors, industries can play a pivotal role in building a resilient and sustainable future.

In "The Role of Industry," the narrative explores the diverse impacts of different sectors on rising sea levels and their efforts to mitigate their environmental footprint. By highlighting the challenges and opportunities within each industry, the chapter underscores the importance of collective action and sustainable practices in addressing one of the most pressing global issues of our time. This discussion sets the stage for exploring the broader implications of intersectoral cooperation and global solidarity in subsequent chapters, further emphasizing the need for a united and inclusive approach to combating rising sea levels.

Chapter 19
Hope on the Horizon

Introduction: A Beacon of Hope

Amidst the daunting challenges posed by rising sea levels, stories of hope and resilience emerge, demonstrating the human spirit's ability to adapt and overcome. This chapter highlights communities that have successfully adapted to changing conditions, innovative solutions showing promise, and global movements pushing for transformative change. These stories provide a sense of optimism and illustrate that, despite the enormity of the crisis, positive change is possible.

Community Adaptation Success Stories

Around the world, communities are implementing effective strategies to adapt to rising sea levels, showcasing ingenuity and resilience.

1. The Netherlands: A History of Water Management

Concept:

The Netherlands has a long history of managing water and has become a global leader in flood control and adaptation strategies.

Examples:

The Delta Works, a series of dams, sluices, locks, dikes, and storm surge barriers, protect the Dutch coast from the sea.

This engineering marvel is considered one of the Seven Wonders of the Modern World.

The Room for the River program creates floodplains and enhances river capacity to manage higher water levels, integrating natural landscapes with urban planning.

"Our approach is to live with water, not fight it. By embracing innovative solutions, we can create a resilient future." - Henk Ovink, Special Envoy for International Water Affairs for the Netherlands.

2. Bangladesh: Community-Based Adaptation

Concept:

Bangladesh, one of the most vulnerable countries to sea level rise, has developed community-based adaptation strategies.

Examples:

Floating gardens, known as "baira," use water hyacinth and bamboo to create floating platforms for agriculture, ensuring food security despite flooding.

Cyclone shelters and early warning systems, combined with community education, have significantly reduced the death toll from extreme weather events.

"Our resilience comes from our ability to adapt and innovate. Through community efforts, we can face the challenges of rising sea levels." - Dr. Saleemul Huq, Director of the International Centre for Climate Change and Development.

3. The Maldives: Embracing Resilient Infrastructure

Concept:

The Maldives is implementing innovative infrastructure projects to combat the impacts of rising sea levels on its low-lying islands.

Examples:

The artificial island of Hulhumalé is being developed with elevated buildings, green spaces, and sustainable urban planning to house displaced residents.

Coral reef restoration projects are enhancing natural barriers against storm surges and coastal erosion.

"We must act now to protect our home. By investing in resilient infrastructure and environmental restoration, we can secure our future." - Mohamed Nasheed, Former President of the Maldives.

Innovative Solutions Showing Promise

Technological advancements and innovative approaches are offering new hope for mitigating the impacts of rising sea levels.

4. Floating Cities: A Vision for the Future

Concept:

Floating cities offer a futuristic solution to rising sea levels, providing habitable spaces on the water.

Examples:

Oceanix City, a prototype designed by BIG (Bjarke Ingels Group), envisions self-sustaining, modular floating communities that can withstand rising sea levels and extreme weather.

The Seasteading Institute advocates for floating platforms that could host communities and businesses, promoting sustainability and resilience.

"Floating cities represent a bold vision for adapting to climate change. By rethinking urban living, we can create sustainable habitats for the future." - Marc Collins Chen, Co-Founder of Oceanix.

5. Blue Carbon: Harnessing Nature's Power

Concept:

Blue carbon refers to carbon captured by coastal and marine ecosystems, such as mangroves, seagrasses, and salt marshes.

Examples:

Restoration of mangrove forests in countries like Indonesia and Kenya is enhancing carbon sequestration, protecting coastlines from erosion, and supporting biodiversity.

Seagrass restoration projects in places like the United Kingdom are reviving underwater meadows, which store significant amounts of carbon and provide habitat for marine life.

"Blue carbon ecosystems are vital for climate resilience. By protecting and restoring these natural habitats, we can mitigate climate change and adapt to its impacts." - Dr. Emily Pidgeon, Senior Director of Conservation International's Blue Carbon Program.

6. Renewable Energy: Powering a Sustainable Future

Concept:

Transitioning to renewable energy sources is crucial for reducing greenhouse gas emissions and mitigating climate change.

Examples:

Offshore wind farms, such as the Hornsea Project in the UK, harness the power of the wind to generate clean energy, reducing reliance on fossil fuels.

Solar energy projects, like Morocco's Noor Ouarzazate complex, are providing renewable power on a large scale, helping to reduce carbon emissions and promote sustainable development.

"Renewable energy is key to addressing climate change. By investing in clean energy solutions, we can reduce our carbon footprint and create a sustainable future." - Dr. Fatih Birol, Executive Director of the International Energy Agency.

Global Movements Pushing for Change

Grassroots movements and international collaborations are driving transformative change and advocating for climate action.

7. Youth Climate Strikes: A Global Call to Action

Concept:

Youth-led climate strikes and movements are demanding urgent action from leaders to address climate change.

Examples:

Fridays for Future, initiated by Greta Thunberg, has mobilized millions of young people worldwide to strike for climate action and push for policies addressing rising sea levels.

Extinction Rebellion uses nonviolent civil disobedience to raise awareness about the climate crisis and demand government action to reduce carbon emissions.

"Our voices matter. We are the future, and we must act now to protect our planet." - Greta Thunberg, Climate Activist.

8. International Agreements: Collaborative Efforts for Climate Action

Concept:

Global agreements and collaborations are essential for coordinated efforts to combat climate change and rising sea levels.

Examples:

The Paris Agreement, adopted in 2015, aims to limit global warming to well below 2 degrees Celsius, with countries committing to reduce emissions and enhance climate resilience.

The Global Climate Action Summit brings together leaders from governments, businesses, and civil society to accelerate climate action and share best practices.

"International cooperation is vital for addressing the global challenge of climate change. By working together, we can achieve a sustainable and resilient future." - António Guterres, United Nations Secretary-General.

9. Indigenous Knowledge: Harnessing Traditional Practices

Concept:

Indigenous communities possess valuable knowledge and practices for sustainable living and climate adaptation.

Examples:

In the Arctic, Indigenous peoples use traditional knowledge to monitor environmental changes and adapt to shifting conditions, enhancing resilience to rising sea levels.

Coastal Indigenous communities in the Pacific Islands employ traditional land management and conservation practices to protect their environments and ensure food security.

"Indigenous knowledge is a powerful tool for climate resilience. By respecting and integrating traditional practices, we can enhance our ability to adapt to a changing world." - Hindou Oumarou Ibrahim, Indigenous Rights Advocate.

Conclusion: A Vision of Resilience and Hope

The chapter concludes by emphasizing that, despite the challenges posed by rising sea levels, there is hope on the horizon. Through community adaptation, innovative solutions, and global movements, humanity is demonstrating its ability to confront the climate crisis with resilience and creativity. By learning from these stories of hope and building on these successes, we can pave the way for a more sustainable and resilient future.

In "Hope on the Horizon," the narrative explores the inspiring efforts of communities, innovators, and activists who are making a difference in the fight against rising sea levels. By highlighting their achievements and the potential for positive change, the chapter underscores the importance of collective action and the human capacity for adaptation. This discussion sets the stage for the final chapter, which will reflect on the lessons learned and the path forward in addressing the global challenge of rising sea levels.

Chapter 20
A Call to Action

Introduction: The Urgency of Now

As the concluding chapter, "A Call to Action" is a rallying cry for readers to engage in the collective effort required to combat rising sea levels. Throughout this book, we have explored the causes, impacts, and innovative responses to this global challenge. Now, we turn our focus to practical steps that individuals, communities, and governments can take to mitigate and adapt to the effects of rising sea levels.

Individual Actions: Empowering Personal Responsibility

Every individual has a role to play in combating rising sea levels. Small actions, when multiplied across millions of people, can create significant change.

1. Reducing Carbon Footprint

Concept:

Personal choices in daily life can reduce greenhouse gas emissions.

Examples:

Energy Efficiency:

Use energy-efficient appliances, LED lighting, and smart thermostats to reduce energy consumption.

Sustainable Transportation:

Opt for public transport, cycling, walking, or electric vehicles to reduce fossil fuel use.

Dietary Choices:

Reduce meat and dairy consumption, as livestock farming is a significant source of methane emissions.

"Our everyday choices matter. By making sustainable decisions, we can reduce our impact on the environment and contribute to the fight against climate change." - Jane Goodall, Environmentalist.

2. Supporting Renewable Energy

Concept:

Transitioning to renewable energy sources at home and advocating for broader adoption.

Examples:

Solar Panels:

Install solar panels on homes to generate clean energy and reduce reliance on fossil fuels.

Green Energy Plans:

Choose utility providers that offer renewable energy options, such as wind or solar power.

"Renewable energy is our path to a sustainable future. By investing in clean energy, we can reduce emissions and protect our planet." - Bill McKibben, Environmentalist and Author.

3. Advocacy and Education

Concept:

Raising awareness and advocating for policies that address climate change.

Examples:

Community Engagement:

Participate in local environmental groups, attend town hall meetings, and support climate action initiatives.

Education:

Educate others about the impacts of rising sea levels and the importance of sustainable practices through social media, community events, and educational programs.

"Education and advocacy are powerful tools. By spreading knowledge and pushing for change, we can influence policies and create a more sustainable world." - Greta Thunberg, Climate Activist.

Community Actions: Building Resilient Communities

Communities can come together to implement collective solutions that enhance resilience and reduce vulnerabilities to rising sea levels.

4. Sustainable Urban Planning

Concept:

Integrating sustainable practices into urban development to create resilient communities.

Examples:

Green Infrastructure:

Develop parks, green roofs, and urban forests to absorb floodwaters, reduce urban heat, and improve air quality.

Flood-Resilient Buildings:

Construct buildings with elevated foundations, permeable pavements, and flood barriers to withstand rising waters.

"Our cities must be designed with resilience in mind. By incorporating green infrastructure and sustainable planning, we can protect communities from the impacts of rising sea levels." - Amanda Burden, Urban Planner.

5. Community-Based Adaptation

Concept:

Implementing localized strategies that leverage community knowledge and resources.

Examples:

Emergency Preparedness:

Develop community emergency plans, conduct drills, and establish communication networks to respond effectively to flooding events.

Ecosystem Restoration:

Engage in projects that restore wetlands, mangroves, and other natural buffers that protect against storm surges and erosion.

"Community-based adaptation is crucial. By working together and utilizing local knowledge, we can build resilience and protect our communities." - Dr. Saleemul Huq, Climate Change Expert.

6. Local Policy Advocacy

Concept:

Advocating for local policies that promote sustainability and climate resilience.

Examples:

Zoning Laws:

Support zoning regulations that restrict development in high-risk coastal areas and promote the preservation of natural habitats.

Infrastructure Investment:

Advocate for investment in resilient infrastructure, such as seawalls, levees, and stormwater management systems.

"Local policies play a vital role in climate resilience. By advocating for sustainable development and infrastructure, we can create safer and more resilient communities." - Dr. Robert Bullard, Environmental Justice Scholar.

Government Actions: Leading the Charge

Governments at all levels must take decisive action to address rising sea levels through policies, regulations, and international cooperation.

7. National Climate Policies

Concept:

Implementing comprehensive climate policies that reduce emissions and enhance resilience.

Examples:

Carbon Pricing:

Introduce carbon pricing mechanisms, such as carbon taxes or cap-and-trade systems, to incentivize emissions reductions.

Renewable Energy Targets:

Set ambitious targets for renewable energy adoption and provide incentives for clean energy investments.

"Effective climate policies are essential. By implementing carbon pricing and supporting renewable energy, governments can drive the transition to a sustainable future." - Dr. Nicholas Stern, Economist.

8. International Cooperation

Concept:

Collaborating with other nations to address the global nature of climate change and rising sea levels.

Examples:

Paris Agreement: Strengthen commitments under the Paris Agreement, aiming to limit global temperature rise and support climate adaptation efforts in vulnerable countries.

Climate Finance:

Increase funding for climate adaptation and mitigation projects in developing nations, ensuring equitable support for those most affected.

"Global cooperation is crucial in the fight against climate change. By working together, we can achieve meaningful progress and protect our planet." - António Guterres, United Nations Secretary-General.

9. Environmental Legislation and Regulation

Concept:

Enacting and enforcing laws that protect the environment and promote sustainable practices.

Examples:

Emission Standards:

Set and enforce stringent emissions standards for industries, vehicles, and power plants to reduce greenhouse gas emissions.

Conservation Laws:

Implement laws that protect natural habitats, promote reforestation, and prevent deforestation.

"Strong environmental laws are the foundation of climate action. By enacting and enforcing regulations, we can safeguard our environment and combat rising sea levels." - Dr. Vandana Shiva, Environmental Activist.

Collective Effort: The Power of Unity

The fight against rising sea levels requires a collective effort, transcending individual, community, and governmental actions. By working together, we can create a more resilient and sustainable future.

10. Public-Private Partnerships

Concept:

Collaborating across sectors to leverage resources and expertise for climate solutions.

Examples:

Green Bonds:

Governments and private companies can issue green bonds to finance sustainable infrastructure projects, such as renewable energy installations and flood defenses.

Corporate Sustainability:

Businesses can partner with governments and NGOs to develop and implement sustainability initiatives, reducing their carbon footprint and promoting environmental stewardship.

"Public-private partnerships are essential for scalable solutions. By collaborating, we can pool resources and drive meaningful climate action." - Al Gore, Former U.S. Vice President and Environmental Advocate.

11. Grassroots Movements

Concept:

Mobilizing communities and individuals to advocate for climate action and hold leaders accountable.

Examples:

Climate Marches:

Organize and participate in climate marches and rallies to demand action from policymakers and raise public awareness.

Petitions and Campaigns:

Launch and support petitions and advocacy campaigns that call for stronger climate policies and corporate responsibility.

"Grassroots movements are the heartbeat of change. By uniting and raising our voices, we can push for the action needed to address rising sea levels." - Naomi Klein, Author and Activist.

Conclusion: A Call to Action

The chapter concludes by reiterating the urgent need for collective action to combat rising sea levels. Each individual, community, and government has a role to play in addressing this global threat. By taking practical steps and working together, we can create a more resilient and sustainable future for generations to come.

In "A Call to Action," the narrative emphasizes that the fight against rising sea levels is not just a challenge but an opportunity for transformative change. By empowering individuals, supporting communities, and demanding action from governments, we can turn the tide and build a better world. This final chapter serves as a powerful reminder that hope lies in our hands, and through collective effort, we can overcome the challenges and protect our planet for future generations.

www.ingramcontent.com/pod-product-compliance
Lightning Source LLC
Chambersburg PA
CBHW071923210526
45479CB00002B/528